A Guide to

Squid, Cuttlefish and Octopuses of Australasia

'Squid' (1996), acrylic on paper by Tatipai Barsa.

A Guide to **Squid, Cuttlefish and Octopuses** of Australasia

Mark Norman &
Amanda Reid

CSIRO PUBLISHING

GOULD LEAGUE

First published 2000

Set in 9.5/11pt Goudy
Print in Australia by Impact Printing

Cataloguing-in-Publication data

Norman, Mark Douglas.
Guide to squid, cuttlefish and octopus of Australasia.

Bibliography.
Includes index.
ISBN 0 643 06577 6.

1. Cuttlefish — Australia — Identification.
2. Cuttlefish — Australasia — Identification.
3. Octopus — Australia — Identification.
4. Octopus — Australasia — Identification.
5. Squids — Australia — Identification.
6. Squids — Australasia — Identification.
7. Cephalopoda — Australasia — Identification.
I. Reid, Amanda Louise. II. Title.

594.50994

Published by:

The Gould League of Australia
Genoa Street
Moorabbin
Victoria 3189
Telephone: +61 3 9532 0909
Fax: +61 3 9532 2860
Email: gould@gould.edu.au
Web site: www.gould.edu.au

CSIRO PUBLISHING
150 Oxford Street
Collingwood
Victoria 3066
Telephone: +61 3 9662 7500
Fax: +61 3 9662 7555
Email: sales@publish.csiro.au
Web site: www.publish.csiro.au

Contents

Acknowledgements

This guide would not have been possible without the help of many people. Special thanks go to David Paul and Julian Finn for their skills, energy and excellent eyes.

The senior author would like to acknowledge the funding support of the Australian Biological Resources Study (ABRS) without which many of the animals here would not have been treated. We would also like to thank ABRS and the Food and Agriculture Organisation of the United Nations (FAO) for permission to use their illustrations. Artists were K. Nolan, K. Hollis, C. Eadie and H. Lewis, their excellent images being produced for Carpenter and Niem (1998) and Lu and Dunning (1998) (see References).

Many thanks to the following people for allowing us to use their great images: R. Kuiter, R. Steene, M. Molloy, C. Bryce, A. Kerstitch, K. Reisenbichler, R. Young, S. Foale, C. Black, A. Lewis, G. Edgar, L. Reynolds, C. Griffith, R. Fenwick, C. Roper, M. Vecchione, A. Semini, Z. Florian, G. Williamson, B. Reid and G. Jackson.

Many thanks to the artist, Tatipai Barsa for permission to include his picture 'Squid', made available courtesy of the Queensland Artworkers Alliance.

The authors would finally like to acknowledge the help of J. Ahern, the staff of Museum Victoria, the Zoology department at Melbourne University, the many volunteers who assisted with field surveys, and the staff of the Marine Discovery Centre, Queenscliff.

Mark Norman **Amanda Reid**

(Photo: M Molloy)

Introduction

The group of animals known as cephalopods includes octopuses and their relatives — squid, cuttlefish and nautiluses. Most of us know little about the behaviour and biology of these creatures. We may see them in fish-markets, eat them in restaurants or occasionally watch them portrayed as 'monsters of the deep' in B-grade horror movies. Often we know them only from their distorted or partial remains washed up on a beach — bones of cuttlefish (commonly used to sharpen the beaks of cage birds) or a coiled Ram's Horn shell. A regular beachcomber may be lucky enough to discover a delicate paper nautilus shell.

In fact, cephalopods are as abundant as fish in the sea and humans have a long history of association with them. Pottery dated to 1500 BC and exquisite mosaics from the first century BC accurately portray octopus and squid found in the Mediterranean Sea. Many coastal cultures throughout the world have included these animals in their diets, daily lives and mythologies — and continue to do so. Calamari (squid) rings and tubes, and pickled or 'baby' octopuses are sold in fish-markets all around the world.

The Australasian region (from southern Asia and New Guinea to Australia and New Zealand) has the greatest diversity of these animals in the world. This book will help you identify the cephalopods most likely to be found in this region. It covers most of the species found in relatively shallow waters, at depths of less than 100 metres. There are many more deeper-water species, and the book gives a few examples of these. Many cuttlebones washed up on beaches are from animals that live in deep water, so pictures of these have been included too.

How much do you know about cephalopods? What is a cuttlebone for? How could one of these soft-bodied animals fashion something as beautiful as a paper nautilus shell? What sort of life did a squid lead before it became a calamari ring? What are blue-ringed octopuses really like? Is a giant squid likely to drag you out of your boat and into the ocean depths?

Whether you are a naturalist, diver, beachcomber, reef-walker or an angler, we hope you find this book interesting and useful.

Sightings of large octopuses or dying giant squid near the water's surface may have led to many of the early sea monster myths attributed to cephalopods.

What are cephalopods?

Cephalopods — together with snails, chitons, clams, mussels and others — are classified together in the phylum Mollusca. The features that link such different-looking animals as squids and limpets include an outside skin-like covering, called a mantle, which is open at one end. The mantle encloses all the major body organs and secretes the external shell of the limpet and the internal shell of the squid. Both have an armed tongue (or radula) and both breathe through feather-like gills in the mantle cavity. They also share the same respiratory pigment, called haemocyanin, in their blood which carries and stores oxygen in the body.

The cephalopods are separated as a distinct group, or class, within the molluscs because they have a ring of arms (which can be considered as 'super-lips') surrounding the mouth.

Evolution of cephalopods

Cephalopods have been around for over 500 million years, once dominating the oceans before the arrival of fish. The early cephalopods were typically protected in a chambered shell, like the chambered nautiluses that still exist today. Some ancient cephalopods reached a massive size, with shells over two metres in diameter.

The oldest group, the nautiloids, dates back to the Upper Cambrian period (500–550 million years ago). This group gave rise to all other cephalopods including many lineages of shelled cephalopods that once dominated ancient oceans but are now extinct.

The ammonites arose in the Devonian period (350–400 million years ago) and had external chambered shells that reached a massive size — more than two metres in diameter.

The belemnites arose in the Carboniferous period (280–350 million years ago), evolving with an internal shell. The belemnites are thought to be the direct ancestors of today's squid, cuttlefish and octopuses.

Ammonites arose in the Devonian period, 350–400 million years ago. (Photo: D Paul)

The fossilised remains of small ammonites. (Photo: D Paul)

Today, the only surviving nautiloids are the few species of chambered nautiluses found in deeper waters off the tropical coral reefs of the Indian and Pacific oceans.

All swimming animals need to be neutrally buoyant so they can remain at a particular depth (neither sink nor float). The extinct ammonites and nautiloids acheved this by means of a chambered and gas-filled external shell, as do the living *Nautilus* and *Allonautilus*. By altering the relative volumes of gas and liquid in the chambers, they could control their buoyancy and therefore their depth in the water column. The shells of nautiloids and ammonites came in many shapes and sizes, including conical, coiled and straight.

While having many advantages, such as providing protection from predators, these shells were sometimes quite unwieldy and prevented these animals from moving quickly. As faster predators, such as the fish, began to inhabit the seas, cephalopods evolved with smaller and lighter shells, allowing them to move more quickly and thus escape more easily.

Over time, the cephalopod mantle enveloped the shell so that the external chambered shell of the straight nautiloids became the internal shell of the belemnites, cuttlefishes and squids. In many groups the shell has been lost altogether.

Loss of the shell also freed cephalopods from depth restrictions, because animals with a gas-filled shell can only survive at depths at which the shell structure is able to withstand external water pressure without imploding. But this meant that some cephalopods had to find alternative ways of controlling their buoyancy.

Some groups, such as the familiar muscular octopuses, evolved to become bottom-dwellers. Others found alternative ways to maintain buoyancy in mid-water — for example, by using a combination of fins and jet propulsion. In some species, the body tissues contain pockets of liquids less dense than seawater (such as ammonia) to lighten them and assist with buoyancy. Some neutrally buoyant species simply drift about with the currents.

The loss of the bulky external shell allowed for the development of a muscular body, which enables jet propulsion, the main form of fast swimming in these animals.

The chambered nautilus is a living relic of the ancient cephalopods. *Nautilus macromphalus* is one of the few species still alive in the world today. (Photo: D Paul)

Cephalopods today

Today, cephalopods are found in all of the world's oceans, from the tropics to the Antarctic and the Arctic, and from shallow intertidal reefs to deep-sea trenches over seven kilometres in depth. All are marine; none survive in fresh water.

They come in all shapes and sizes, ranging from a pygmy squid, which would fit on your thumbnail, to the Giant Squid, *Architeuthis*, which can reach almost 20 metres in length and weigh 250 kilograms.

All cephalopods have the same basic body plan. A set of limbs comes off the head in a ring around the mouth. Most other major body parts, the digestive organs, reproductive and excretory systems and the gills, are contained in a muscular bag (the mantle) at the back (posterior) of the animal.

Chambered nautiluses have about 90 suckerless tentacles around the mouth. All other living cephalopods have eight arms which possess suckers and/or hooks. The arms are numbered as pairs, one to four, from the upper surface to the lower side.

Squid and cuttlefish differ from octopuses by having an additional pair of limbs known as feeding tentacles. These tentacles emerge from between the bases of arms three and four, with many groups having pouches into which these limbs can be retracted. The long feeding tentacles are typically used to shoot out and grab prey.

The mantle forms a cavity through which water passes, carrying oxygen to the gills. The creature is able to eject water from the mantle via a tubular funnel, giving it the capacity for jet propulsion. When the mantle is relaxed, water goes into the mantle cavity. When the mantle contracts, the edge is sealed and water is forced out through the opening in the funnel. The funnel can be pointed in different directions, enabling both forward and backward movement. Rapid contractions of the muscular mantle forces water at considerable pressure through the funnel, enabling the animal to jet rapidly away from a predator, launch an attack on its prey, or even squirt potential predators in the face — as some shore-dwelling octopuses do to herons!

Some species can also move about by undulating their body fins ('fin swimming'). Species with

Cephalopods have a muscular 'mantle' which houses their main organs. They propel themselves by expelling water through the funnel. (Photo: D Paul)

Octopuses such as the Southern Blue-ringed Octopus can swim considerable distances using jet propulsion, trailing their arms behind them. (Photo: M Norman)

webs between the arms can also swim by spreading, then drawing their arms together, rather like opening and closing an umbrella. This is called 'medusoid' swimming, named after the swimming motion of jellyfishes.

A distinctive feature of cephalopods is their ability to change colour and texture. They are the chameleons of the sea. They perform colour changes using small special organs in the skin known as 'chromatophores'. These are tiny elastic bags of dense pigment that can be expanded and contracted to show a spot of a particular colour, a bit like the pixels in a television screen. Turning these spots on and off allows them to display particular colours or patterns.

Chromatophores come in up to five colours and can be present in densities of more than 200 in a square millimetre. The skin of most cephalopods also contains a layer of reflective tissue producing an iridescent sheen of blue or green.

Above: A Giant Cuttlefish in camouflage mode. Many cephalopods can change the colour and texture of their skin to mimic their surroundings. (Photo: M Norman)

Below: A Bigfin Reef Squid showing the chromatophores and iridescent tissue which produce their dramatic colour pattern. (Photo: C Black)

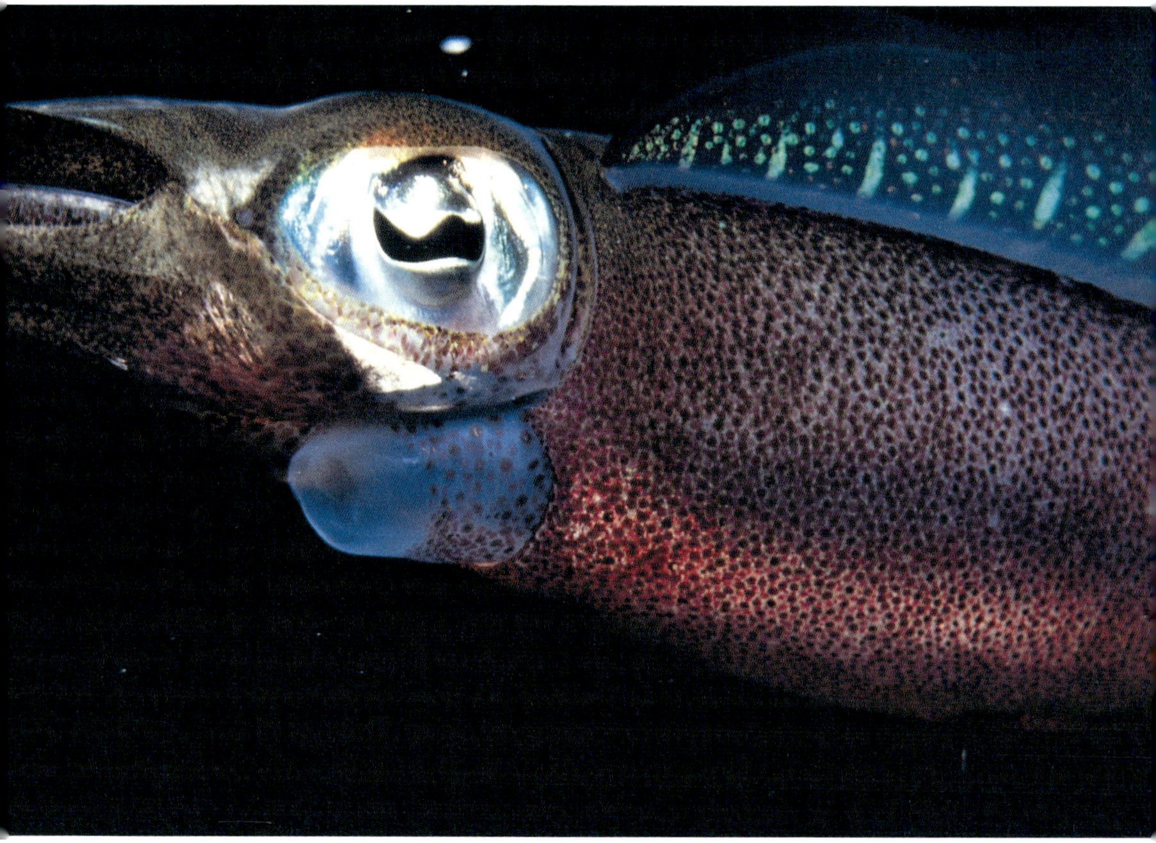

These colours are produced by reflective crystals in structures known as iridocytes. The dazzling blue markings of the blue-ringed octopuses are produced by concentrated iridocytes.

Many cephalopods are capable of changing the texture of their skin to mimic their surroundings. Most cuttlefish and octopuses can push up flaps of skin to match the shape of seaweeds or corals. These bumps are known as papillae and are formed by the contraction of rings of muscle around their base, causing the flaps of skin to stand up.

Cephalopods are considered to be the most intelligent of all the invertebrates (animals without backbones). With a highly developed nervous system and centralised brain, they are capable of complex visual communication and diverse behaviours. They can even learn and solve problems.

Like us, they have very well-developed eyes consisting of a lens and a large retina. Strangely though, for creatures that have acute vision and are themselves so brightly coloured, most cephalopods are colour-blind. They appear to respond to differences in light intensity and polarity (the alignment of light waves), rather than differences in wavelength (which we interpret as colour).

The eye of a Giant Cuttlefish. Although colour-blind, cephalopods have well-developed eyes and acute vision. (Photo: M Norman)

A Hammer Octopus darkens its eyes to appear bigger to potential predators. (Photo: D Paul)

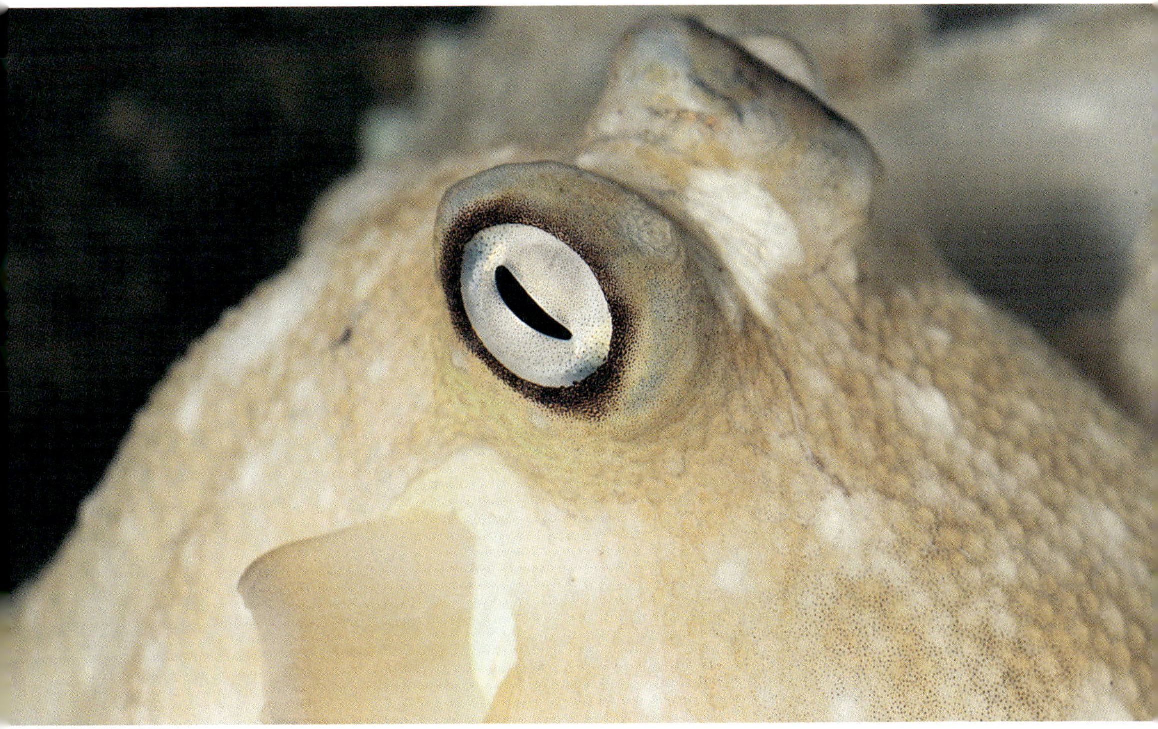

Many octopuses use their toothed tongue (the 'radula') to drill through shells in order to poison and devour their occupants. (Photo: M Norman)

A Southern Calamari Squid devouring a freshly caught fish. (Photo: M Molloy)

All cephalopods are carnivorous. Most are active predators, feeding mainly on fish and crustaceans, as well as other molluscs and marine worms. They grow very fast — often from a few milligrams to a kilogram or more in a year and are generally thought to be short-lived, with life spans ranging from a few months to two or three years, though some large or cold-water species may live longer. Most species of octopus and some squid die after spawning, but other cephalopods may spawn several times.

These creatures are well equipped to catch and eat their prey. They have strong suckers on their limbs that work like suction cups for gripping. In some squid and cuttlefish the rims of the suckers are armed with horny, barbed teeth. In others, the suckers are replaced by scythe-like hooks that can snare prey.

Squid and cuttlefish shoot out two feeding tentacles to first grab the prey and then bring it back toward the ring of arms surrounding the mouth where they hold and manipulate it. Cephalopods' mouths have a hard beak, resembling a parrot's beak, which they use to kill or paralyse prey by injecting poisonous saliva. They use their beak to cut the prey into pieces, then

grind the food into a mush with their tongue, which is armed with rows of sharp teeth.

Some octopuses use this armed tongue (the radula) to drill through the shells of bivalves such as mussels or clams. They inject poisonous saliva through the hole to relax the shellfish, then use their arms to pull apart the two halves of the shell to eat its fleshy occupant.

The radula is one of the links between cephalopods and other molluscs, such as snails, slugs and shellfish. Most molluscs have a toothed radula — even the common garden snail has a radula for scraping up plant matter.

Cephalopods have a range of defences that they use to protect themselves from their many predators. They can change their body colour and pattern as camouflage; they squirt ink as a visual or chemical decoy; they bury themselves in the sandy seafloor or form a sand coat to hide under. Some use poisons, others produce light to cancel their silhouette from below, and some even drop arms or other body parts as decoys.

Reproduction varies considerably between different cephalopod groups. Some have complex courtship behaviour and displays prior to mating. Many cuttlefish and shallow-water squid have spectacular colour changes associated with courting and mate-guarding, including dramatic display combats between rival males. Courtship in octopuses can include displays of specially enlarged suckers on the arms, possibly advertising their readiness to mate. However, the courtship behaviour of most cephalopods is still relatively unknown.

Male Giant Cuttlefish pulse with moving stripes along their body when displaying to females or chasing off rivals. (Photo: M Norman)

When mating, the male Blue-ringed Octopus shunts sperm packages along its third right arm into the female's oviduct. (Photo: M Norman)

Below: The eggs of a Striped Pyjama Squid. (Photo: M Norman)

Mating for all species consists of the male passing packets of sperm (spermatophores) to the female, typically using one or two specially modified arms (known as the 'hectocotylus'). In a few species, the penis is used to directly transfer these sperm packets. Females of many species store the sperm in special receptacles. The location of sperm storage varies between species, from inside the oviducts and ovary, to elsewhere in the mantle cavity, on the head, under the skin on the body or arms, or in a special pouch below the beak. The sperm is either used immediately to fertilise eggs or can be stored for long periods, in some species for up to ten months.

Fertilisation takes place externally in all cephalopods except octopuses, where the eggs are fertilised within the oviducts or ovary. Nautiluses, most squid and all cuttlefish attach their eggs to the seafloor, typically on or under hard surfaces. Some open-ocean squid release hundreds of thousands of tiny eggs in large, transparent, jelly-like masses which float midwater. All these species release their eggs then leave them, most dying after a single spawning event.

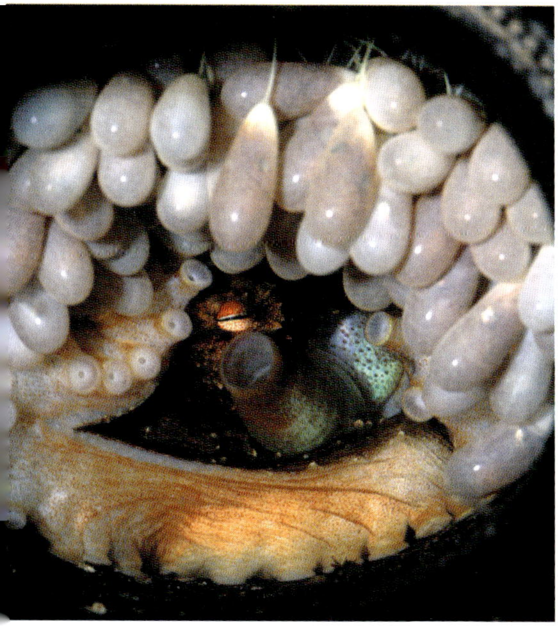

A female Southern Keeled Octopus brooding its eggs. (Photo: D Paul)

Octopuses differ in that females brood their eggs, tending and cleaning the eggs until they hatch. The adult female then dies. Most octopuses lay their eggs in crevices or lairs, sometimes in shells. The Blue-ringed octopus carries egg strings in its web. Even pelagic octopuses, which spend their whole lives in the open ocean, brood their young. Argonaut octopuses brood egg strings within a special egg case (the 'paper nautilus' shell), blanket octopuses attach eggs to small secreted rods, while the 'football' octopus broods the eggs within greatly elongated oviducts.

A recently hatched Southern Keeled Octopus. Young cephalopods are immediately independent. (Photo: D Paul)

The young develop without any parental assistance. The tiny planktonic hatchlings of many species are often carried large distances in surface ocean currents. The young of other species, including some bottletail squids and cuttlefishes, and many octopuses, are larger and well-developed. On hatching, they rapidly settle on the seafloor to adopt an adult-like lifestyle.

A Southern Dumpling Squid hides itself from predators during the day. It uses two arms to rake sand over its body. (Photo: M Norman)

While our knowledge of cephalopods is improving, there is much yet to learn. Scientists continue to discover and describe many new species (particularly in Australian and nearby waters), and are still learning about their distribution in the oceans. Very little, however, is known about the ecology and general biology of most species in this fascinating group.

The main cephalopod groups

Chambered Nautiluses

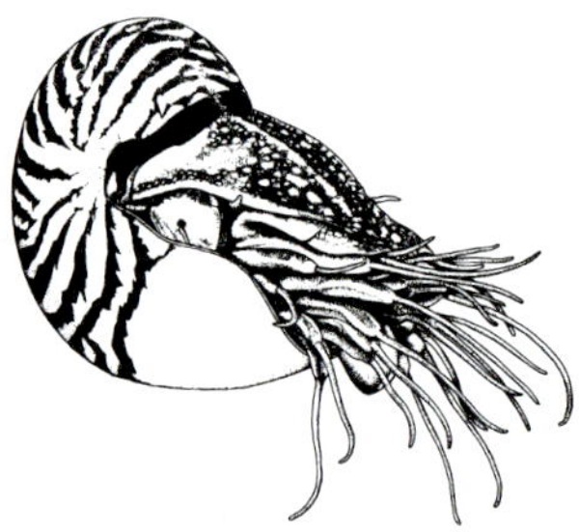

The subclass **Nautiloidea** is represented today by the chambered nautiluses, which have kept the external shell characteristic of their ancestors. The shell is divided into more than 30 chambers connected by a tube. The animal lives in the outermost (most recently made) chamber which can be closed by a leathery hood, known as the operculum. Its hard chalky tipped beak is enclosed within a circle of about 90 retractable arms, which have no suckers or hooks. The eyes are simple and do not have a lens or a cornea. Water flows in and out of the eye which works like a pinhole camera. Underneath the head, a funnel is formed from two overlapping muscular flaps and there are four gills at the back of the mantle cavity, but no ink sac. Nautiluses are mainly scavengers and have a life span of about 20 years.

All the rest of the living cephalopods fall into the subclass **Coleoidea**. They have two gills in the mantle cavity, eight arms (plus two retractable tentacles in squid and cuttlefish), and well-developed eyes with a cornea and a lens. If a shell is present it is inside the mantle of the animal.

Cuttlefish, Bobtail Squid and their allies

This group includes bobtail squid, Ram's Horn squid, cuttlefish, pygmy squid and bottletail squid. All members of this group have eight arms and two feeding tentacles. The cuttlefishes have an internal chambered shell, called the cuttlebone, which is divided into thin flattened chambers and is used for controlling buoyancy. The shell is coiled in the Ram's Horn squid, while in other members of the group the shell (if present at all), is reduced to a thin rod, which gives some support to the body but does not provide buoyancy.

The mantle is very muscular and the funnel is a complete tube. The arms have suckers all the way along their inner sides and these suckers sit on raised, muscular stalks. The feeding tentacles are smooth along most of their length and have a pad of suckers, called the club, at their tip. These tentacles can be retracted into pockets on either side of the mouth. Many members of this group have suckers with a horny rim, which is usually serrated with sharp 'teeth'.

The body is flattened between the upper and lower sides and shield-shaped in cuttlefish but more rounded and globular in the bobtail and bottletail squids. The mantle has many pigment cells which can produce a vast array of colours and patterns for mating displays and camouflage. Most species have a sac containing ink which can be squirted into the water when the animal is irritated or frightened. Bottletail squid, bobtail squid and many cuttlefish can bury themselves in sand so that only their eyes protrude. In this way, they can hide from predators and ambush unsuspecting prawns or crabs. Many emerge at night to feed in open water. Swimming is achieved by the use of the funnel and the fins. In cuttlefish, the fins form a narrow flap around the outer edges of the mantle, while the other members of this group have rounded fins which occupy a smaller proportion of the mantle.

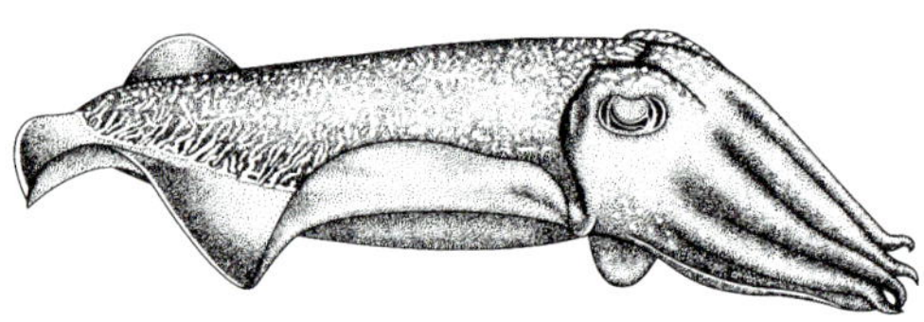

Squid

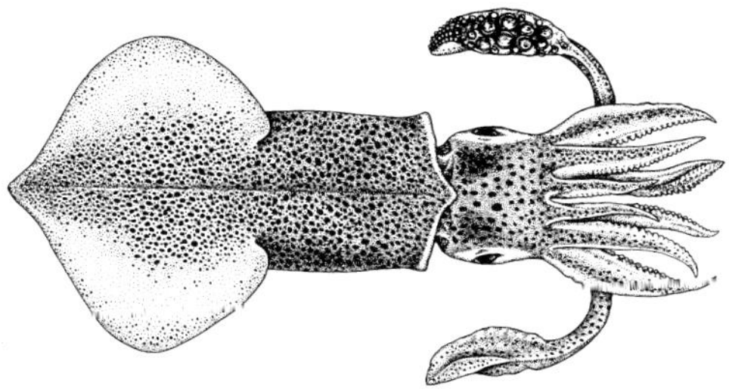

This group includes many families, from the muscular fast-swimming squids (such as the pencil or arrow squids), to open-ocean luminescent and transparent squids (jewel and glass squids), to the biggest invertebrate in the world — the giant squid.

Squid, like cuttlefish and their relatives, have eight arms and two tentacles. The main difference between squid and cuttlefish is that squid lack a broad, chambered shell. The shell is replaced in many species by a horny rod, called a gladius. As a result, squid tend to be cylindrical and torpedo-shaped. The fins of squid are usually confined to the posterior end of the body and are often triangular or arrow-shaped. The suckers are usually confined to the inner sides of the arms and to the clubs of the tentacles.

Some species, such as the Giant Squid, *Architeuthis*, also have suckers scattered along the length of the tentacles and some squid have extra suckers on the membrane surrounding the mouth and beak. Sometimes the suckers are replaced by scythe-like hooks.

Squid eat a variety of foods, but most hunt and eat fish, killing them by biting through the spinal cord rather than using poison. Sometimes they behave as if in a feeding frenzy among a school of fish, killing far more than they can eat.

Vampire Squid

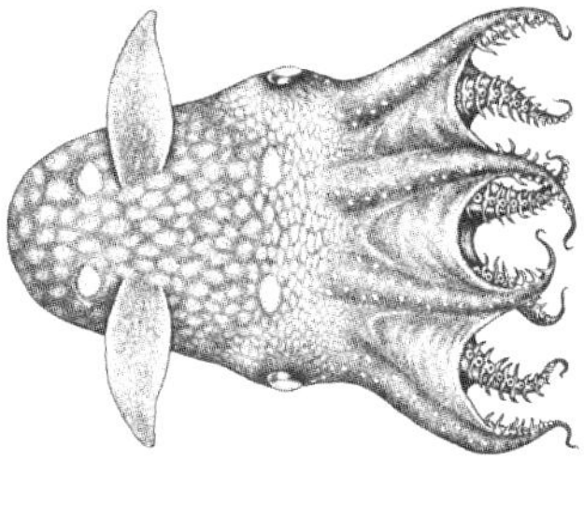

There is only a single species of the distinctive vampire squid which lives between one and three kilometres depth in tropical and temperate waters. It has the consistency of a jellyfish and is maroon to black in colour. The eight arms are joined by deep webs that extend almost to the arm tips, giving it the appearance of an umbrella. It also has a pair of reduced filament-like arms which are stored in pockets between the first and second arms (counting from the upper side of the animal).

Octopuses

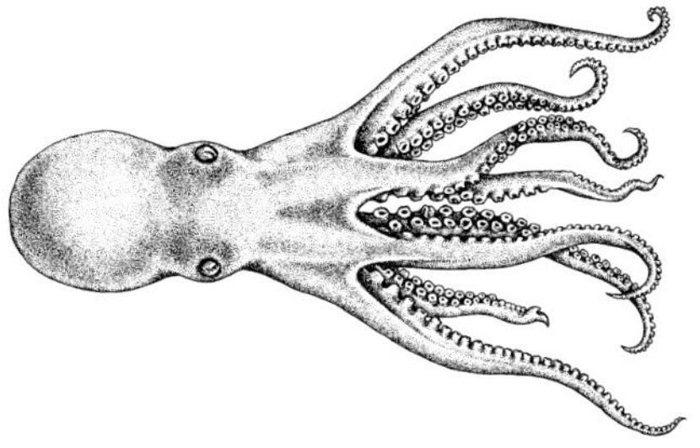

Octopuses have eight arms and no tentacles. Only deep-sea gelatinous species have fins on the body. They have no shell, although some groups possess a pair of small keratin rods in the mantle. Bottom-dwelling octopuses often live in lairs (either permanent or temporary) from which they make feeding expeditions. The suckers, unlike those of squid and cuttlefish, are not raised on muscular stalks and do not have a horny toothed rim.

In contrast to squid, most octopuses crawl over the seafloor but they can also travel using jet propulsion from the funnel, with the arms trailing behind. Octopuses hunt a wide range of prey, including crabs, shellfish and even other octopuses. Once caught, their prey has little chance of escaping because of the strong adhesion of the octopuses' suckers. Most octopuses use the beak and radula to inject toxic saliva, which quickly paralyses the prey. Octopuses have a wide range of activity patterns and behaviours. Most octopuses emerge at night to feed. Some only feed on intertidal reefs while the tide is out.

Octopuses occur in all oceans of the world, from intertidal reefs to depths of more than seven kilometres. Some groups, such as the telescope octopuses, glass octopuses and argonauts, spend their entire lives without touching the seafloor, feeding and living in the water column of the open ocean.

Chambered Nautilus *Nautilus pompilius*

Above: The Chambered Nautilus has a thick leathery hood which it uses to seal off its shell when attacked. (Photo: R Kuiter) *Below:* Cross-section of nautilus shell showing chambers with remnants of gas tube joining chambers. (Photo: D Paul)

Nautiluses live in deep water (to about 500 metres) during the day, and move into shallower water to feed at night. Their shells reach about 18 cm across. Sensory cells on the inner surface of their numerous sucker-less tentacles are used to find food, usually the bodies of dead animals. They adjust their buoyancy by altering the amounts of fluid and gas in the chambers of the shell. Overfishing is placing a severe threat on this species. They are trapped for the shell trade at night. Other species found in the region are *Allonautilus scrobiculatus*, *Nautilus macromphalus* and *Nautilus stenomphalus*.

IDENTIFICATION Central core, or depression, of shell (umbilicus) is small and partially covered by thickening (callus). In other species in the region the umbilicus is open and not covered with a callus (see page 10).

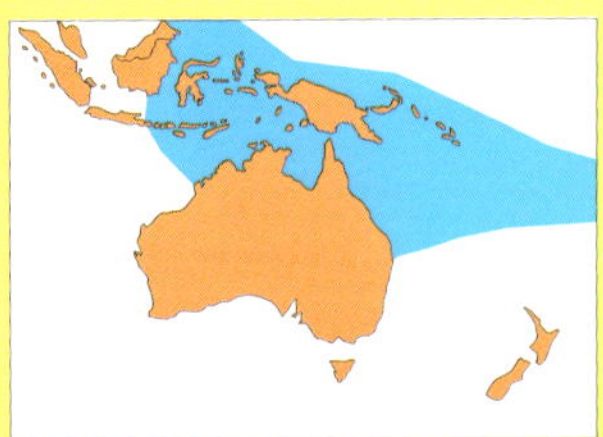

Ram's Horn Squid *Spirula spirula*

These short, cylindrical squid (about 6 cm long) have a spirally coiled internal shell in the posterior end of the body. During the day they concentrate at depths of 600–700 metres, moving into shallower water (about 300 metres) at night. They have a large light organ (photophore) between their rounded posterior fins, the function of which is unknown. These squids are able to control their buoyancy, or ability to float at particular depths, by adjusting the volume of gas and liquid in the internal chambers of the shell. They must float head down in the water. Young animals have been collected at depths of about 1000–1750 metres, suggesting that females probably lay eggs at the bottom of the continental slope. At these depths, the pressure on the shell is more than half a tonne.

The two lower arms of the male have strangely spiked tips, used to transfer spermatophores to the female.

Above: The edge of internal coiled shell is just visible in this animal. *Inset:* The shell of this squid is often found washed up on beaches. Whole animals are rarely seen. (Photos: D Paul)

IDENTIFICATION
Spirally coiled internal shell.

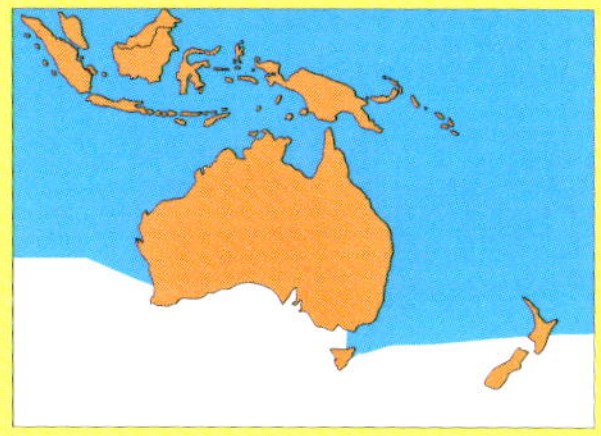

Pfeffer's Flamboyant Cuttlefish *Metasepia pfefferi*

(Photos: R Steene, *top*; CFE Roper, *above*)

This beautiful cuttlefish has a squat, robust body, and shows striking colour patterns often accompanied by a display of raised flap-like papillae and flattened arms. The prominent white bars on the mantle are characteristic of this species. It is a smallish cuttlefish, reaching a mantle length of about 8 cm. The cuttlebone is an unusual diamond shape, yellowish in colour and lacks a spine. It is very pointed at the posterior end and is quite broad when viewed side-on.

Metasepia pfefferi has been observed ambling along the seafloor during the day, displaying its striking colour pattern. This cuttlefish lays round white eggs (approximately 1 cm across) which are attached singly to hard surfaces.

IDENTIFICATION Striking chocolate brown and white colour pattern, with pair of prominent white bars on mantle. Cuttlebone diamond-shaped and yellowish in colour.

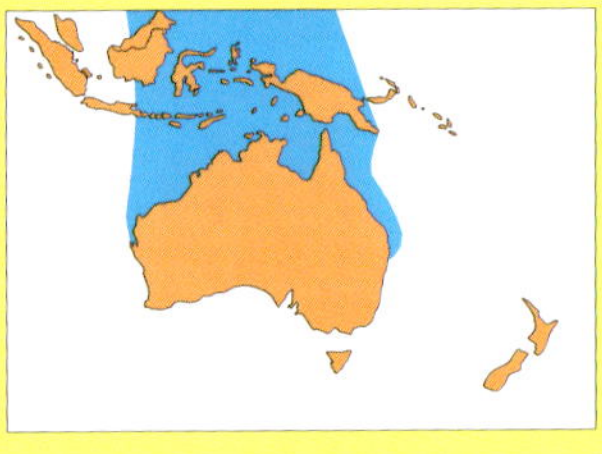

Giant Cuttlefish *Sepia apama*

This is the largest of the cuttlefish, with a mantle length reaching up to half a metre and a total length of up to one metre. It is often encountered by divers around eastern and southern Australia. Like other cuttlefish, it is expert at colour change and camouflage. Colour pattern also plays an important part in communication, particularly in the breeding season. After spawning most cuttlefish die. Bones of the spent animals wash up on beaches in large numbers, many showing the teeth marks of dolphins, birds and fish which feed on the dead and dying animals.

Above: Sepia apama is excellent at raising flaps of skin over its body to impersonate seaweed fronds, both to hide from predators and to sneak up on prey. (Photo: J Finn)
Below: Males show rippling stripes along their body when guarding a chosen female from other males. They also extend and flatten the webs of their arms to give an impression of large overall size. (Photo: M Norman)

IDENTIFICATION Twin rows of three flap-like papillae over each eye. Bones of large animals lack spine and have rough V-shaped thickening (callus) at posterior end. Outer border of bone very wide posteriorly and flared in adults.

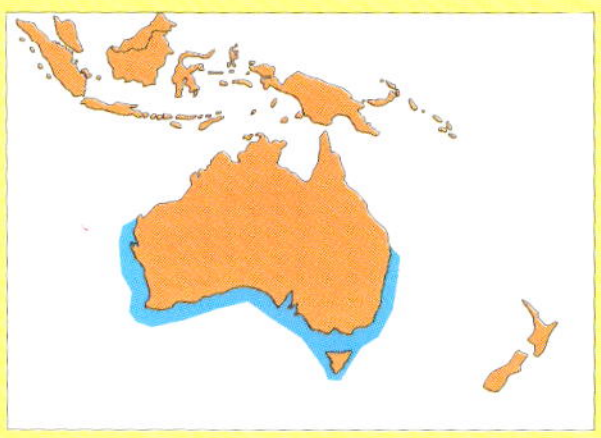

Stumpy Spined Cuttlefish *Sepia bandensis*

Right: An adult showing the pair of lower arms used to amble along the seafloor when foraging at night. *Below:* Same animal with longitudinal skin ridges raised as flaps on body. (Photos: M Norman)

This small cuttlefish reaches a size of about 5 cm in mantle length. It is often seen at night 'ambling'— walking along the seafloor using a lower pair of arms and raised fingers of skin on the underneath of the body. It forages over sand, and amongst algae, coral rubble and live coral. This species tends to seek cover in crevices rather than jetting away from potential predators. Colour patterns range from white through mottled patterns to uniform dark chocolate brown. The latter colour is used when it hides against the sides of black sea cucumbers. *Sepia bandensis* is not well known and may be confused with other small species in the region.

IDENTIFICATION

Longitudinal skin ridges on upper body. Iridescent blue-green spots along base of fins. Cuttlebone very broad oval, upper side sculptured in reticulate pattern. Spine reduced to small stump. Tentacle clubs short with 5 suckers in transverse rows.

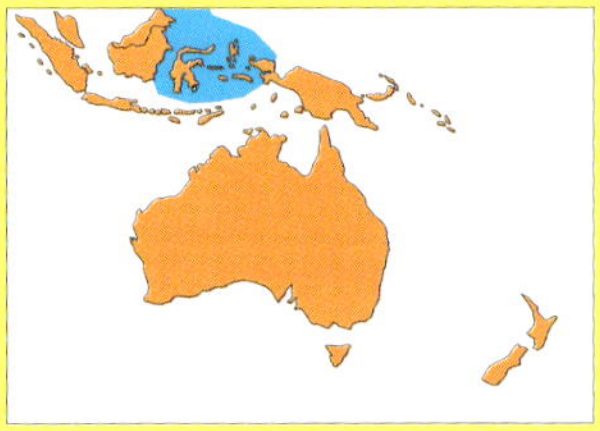

Broadclub Cuttlefish *Sepia latimanus*

Above: A female Broadclub Cuttlefish about to lay eggs. *Below:* A courting male. (Photos: M Norman)

This species is common in the tropics through south-east Asia and northern Australia to depths of at least 30 metres. It is a large cuttlefish, up to about 40 cm in mantle length and weighing up to 5 kilograms. In many places it is actively fished as food and, along with *Sepia pharaonis*, is often sold gutted and dried. It is often seen by divers around coral reefs. Its common name refers to the large flat tentacular clubs.

Females lay white eggs the size of ping-pong balls which are pushed individually into gaps in staghorn corals where they harden and become impossible to extract. Males often guard their mates, attending the female as she lays, and fend off other males.

IDENTIFICATION Tentacular clubs with 5-6 suckers in transverse rows; suckers not uniform in size. Cuttlebone with thick narrow ridge at the posterior end.

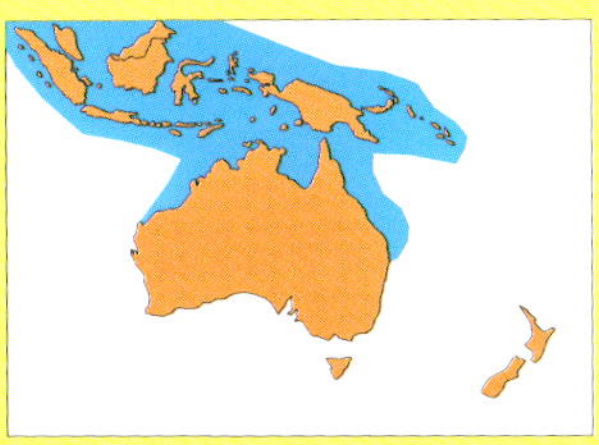

Kisslip Cuttlefish *Sepia lycidas*

Above: A courting male. (Photo: G Williamson) *Below:* The cuttlebone of *Sepia lycidas*. (Drawing: K Hollis/FAO)

IDENTIFICATION Tentacular clubs with eight similar-sized suckers in tranverse rows. Lower side of cuttlebone with deep V-shaped depression, posterior end with thick, rounded ridge.

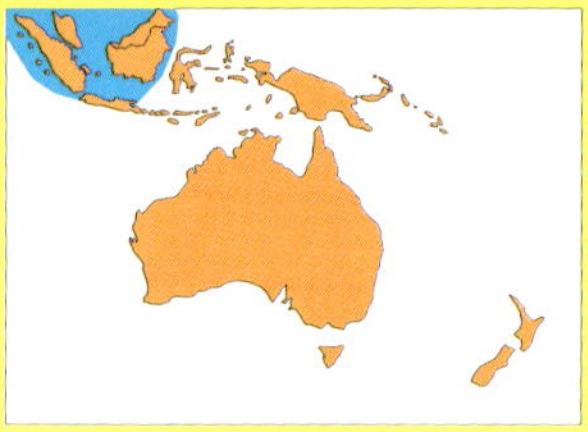

This species is an important component of the commercial cuttlefish fishery in Thailand and Vietnam. It is quite a large animal, reaching about 40 cm in mantle length, and can weigh up to 5 kilograms. It is found at depths to at least 100 metres, moving into the shallows to spawn. The mantle flesh is thick and tasty, making it highly regarded for human consumption. The mouth-like patterning on the body of *Sepia lycidas* has led to its common name, Kisslip Cuttlefish.

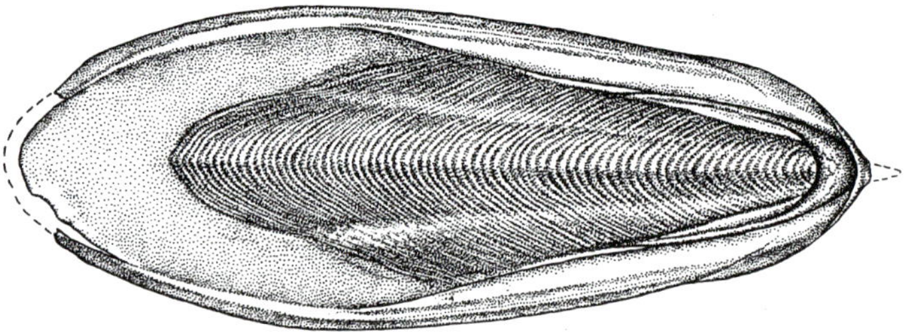

Reaper Cuttlefish *Sepia mestus*

This cuttlefish reaches about 14 cm in mantle length. It occurs on reefs in shallow warm temperate waters of eastern Australia. It is often seen by divers. There have been reports of this species from northern China and Vietnam but it unlikely that these northern animals are indeed the same species. While capable of showing diverse colour patterns, this cuttlefish is often seen to turn a deep rich red colour and is sometimes called the 'red cuttle'. The bone, which is quite flattened and oval in shape, is often found on beaches, particularly around Sydney.

Above: The Reaper Cuttlefish often hides amongst sea urchins in rock crevices. (Photo: R Kuiter) *Below:* A pair showing their squat body shape and short arms. (Photo: R Steene)

IDENTIFICATION Typically red with pair of dark, round spots on upper side of mantle. Cuttlebone oval in outline with rounded ends. Upper side of bone smooth and white.

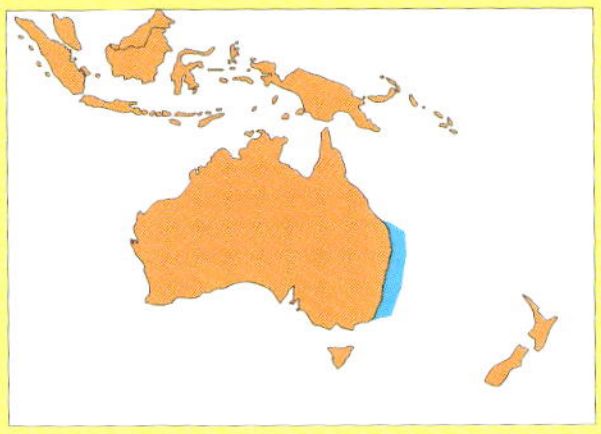

Papuan Cuttlefish *Sepia papuensis*

The Papuan Cuttlefish reaches a maximum size of about 11 cm in mantle length. It has been found in areas of silt, sand or muddy substrates at depths between 13 and 150 metres. It is regularly taken as bycatch of prawn and mixed species trawl fisheries in the Gulf of Carpentaria, Australia. Like other cuttlefishes, this species is capable of marked colour changes, with the ability to create a variety of mottled patterns on the body, often with a pair of characteristic dark 'eye spots' appearing on the posterior end. It is one of the few species in the Australasian region in which the pattern, texture and behaviour has been studied in detail.

IDENTIFICATION Tentacular clubs with 5 to 6 suckers in transverse rows. Surface of head covered with small bumps (papillae). Arms 1–3 have wide membranes on either side of suckers.

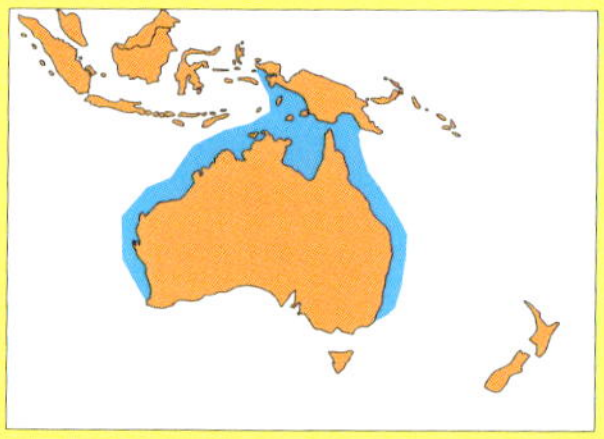

(Photo: M. Norman *top;* CFE Roper *above*)

Pharaoh's Cuttlefish *Sepia pharaonis*

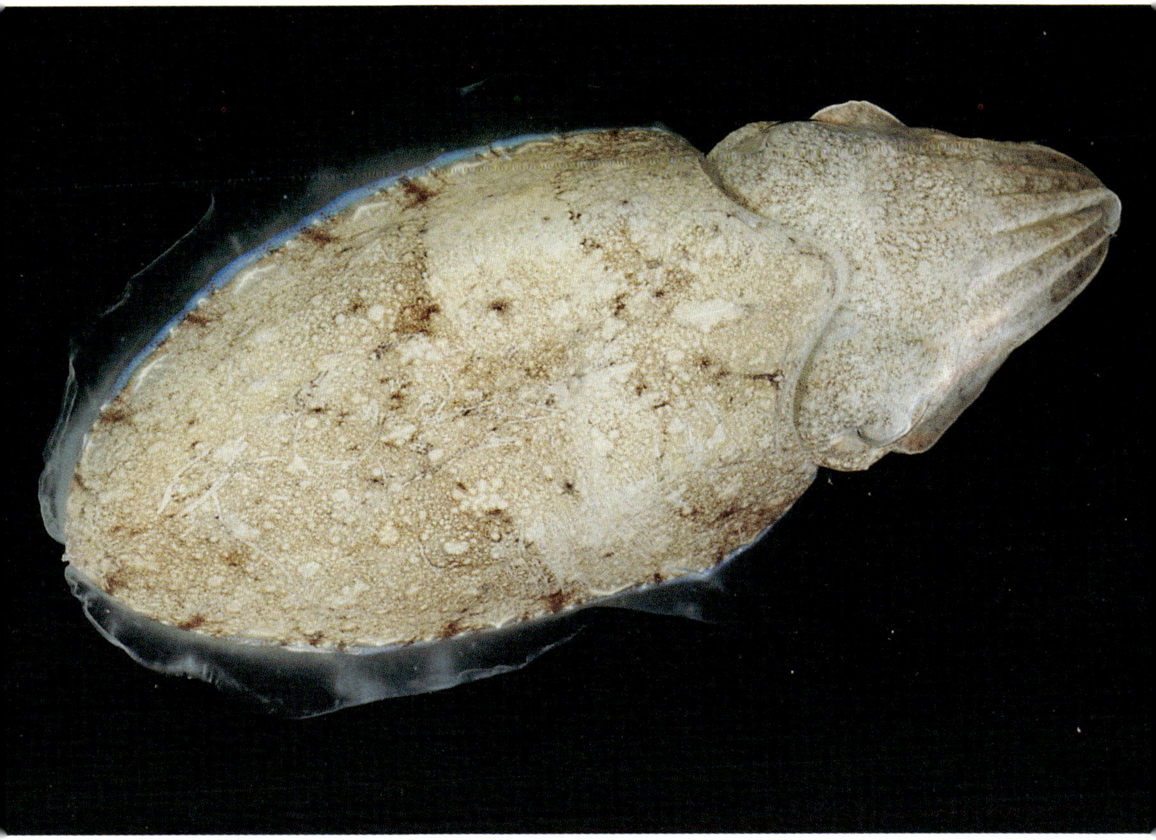

Sepia pharaonis has irregular small pale spots on the mantle, including a pair of asterisk-shaped markings across the middle of the mantle. (Photo: C Bryce)

This large cuttlefish can weigh up to 5 kg and its mantle reaches about 40 cm in length. It is found in shallow coastal waters to about 100 m depth. Pharaoh's Cuttlefish has thick, tender flesh and is prized for human consumption, being an important fisheries species in the Australasian region. Trap fishermen in Hong Kong used to tether a female to attract and trap males. Little is known of its biology.

During the mating season this animal puts on a spectacular display of vivid stripes on the head, arms and body. The pattern is much more pronounced in males than in females.

IDENTIFICATION

Posterior end of cuttlebone has distinctive shiny, bulbous swelling. Pale reflective line along base of each fin.

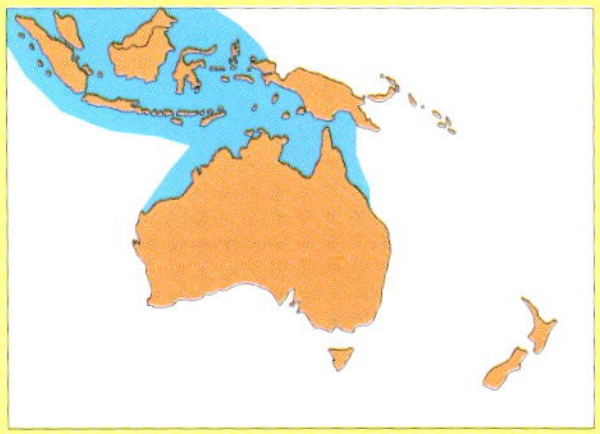

Mourning Cuttlefish *Sepia plangon*

Above: Sepia plangon is frequently encountered in seagrass beds in shallow waters. *Below:* In some colour patterns, it can display numerous fluorescent green wavy bands across the upper mantle. (Photos: M Norman)

This medium-sized cuttlefish, reaching about 15 cm in length, gained its common name for the appearance of the eyes which are often surrounded by a bluish tinge. It is capable of closely mimicking the texture of seaweed and other substrates by erecting the body papillae and flattening the webs on its arms. The Mourning Cuttlefish is frequently observed settled on the seafloor in a tripod position, using the two lower arms to raise the head. It has been collected at depths to about 80 metres, often as bycatch in trawl fisheries and is used for human consumption and as bait.

IDENTIFICATION Colour pattern highly variable. Like *Sepia latimanus*, often seen with two narrow black bands across mantle. Two large dark spots towards posterior end of mantle often displayed. Distinctive deep groove down centre of lower surface of cuttlebone. Posterior part of bone curved to form broad ledge.

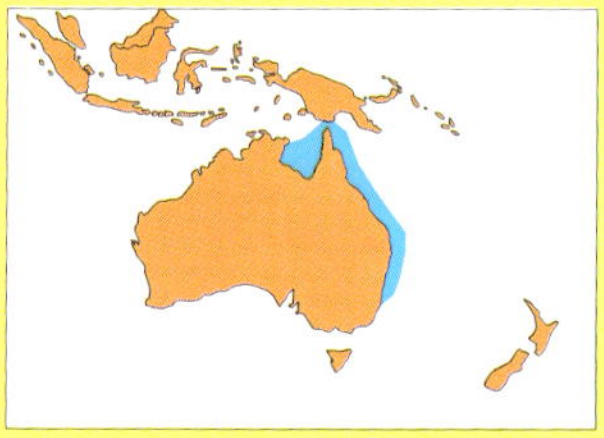

Smith's Cuttlefish *Sepia smithi*

Sepia smithi is a medium-size cuttlefish, up to 17 cm in mantle length. It is found in northern Australia from Shark Bay in the west to Moreton Bay, Queensland, as well as in Timor and the Arafura and Coral Seas. It is often confused with *Sepia whitleyana*. Live animals have been encountered on shallow sandy areas in association with seagrass beds. They often partially bury themselves in sand, their pale colour matching the sea floor. Little is known of their biology.

IDENTIFICATION Tentacle clubs with 20 minute suckers in transverse rows. Posterior end of cuttlebone has distinct yellowish ridge.

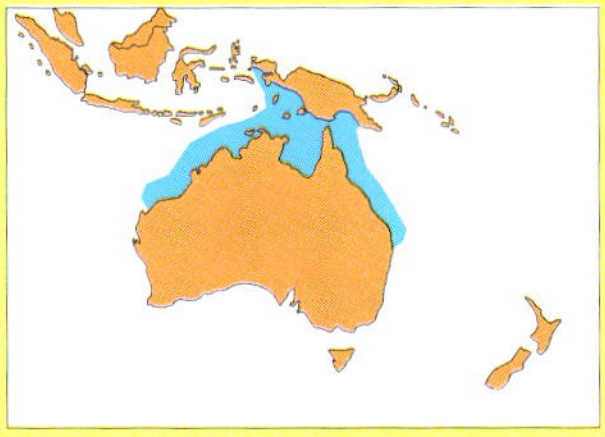

(Photos: M Norman)

Spineless Cuttlefish *Sepiella inermis*

Right: *Sepiella inermis* has 8–9 iridescent gold spots along the base of each fin. *Below:* The function of the dark secretory gland in the posterior tip of its mantle is unknown. (Photos: M Norman)

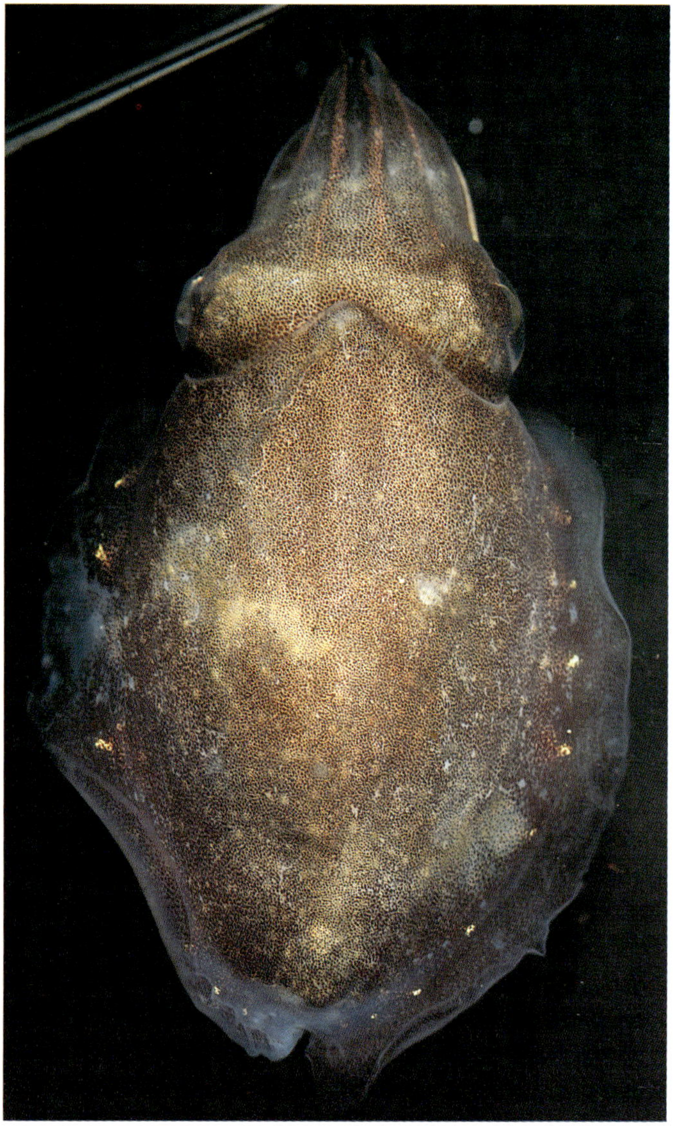

IDENTIFICATION Easily identified by gland in posterior tip of mantle, flanged cuttlebone and regular iridescent spots along upper base of each fin. Tentacle clubs with 12–24 suckers in transverse rows. Colour patterns vary from dark mottled brown to cream with a purple tinge to the fins.

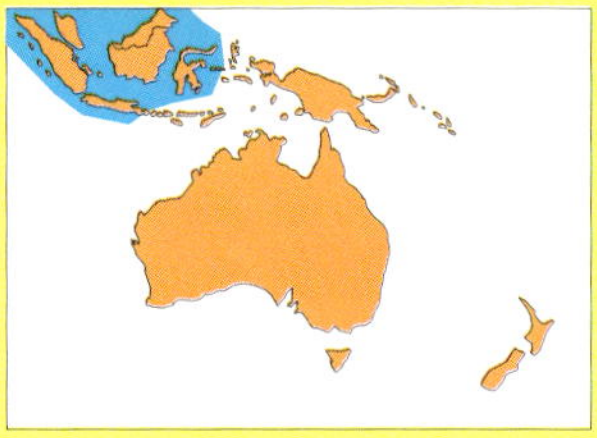

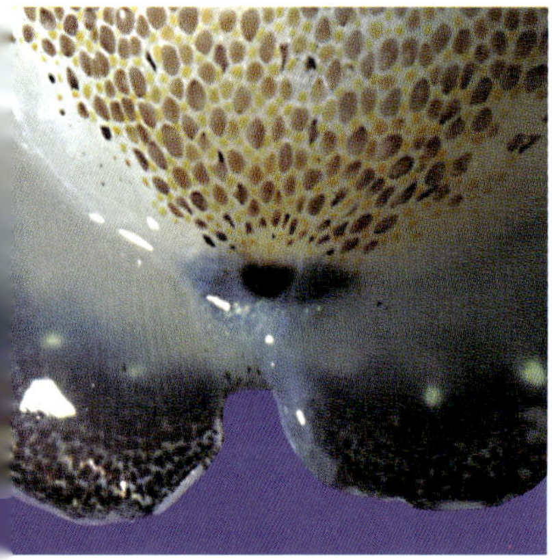

Sepiella inermis is a coastal species, reported to depths of about 40 m, occuring in the northern Indian Ocean, and east to the South China Sea. It reaches about 13 cm long and forms the basis of major commercial fisheries in India and Sri Lanka. A related species, *Sepiella weberi*, occurs in Indonesia and off north-west Australia.

Members of the genus *Sepiella* are the only cuttlefishes to possess a dark secretory gland at the posterior end of the body between the fins. The cuttlebones of *Sepiella* also differ from those of other cuttlefishes in that the posterior end has a thin, flat rim, rather than being thick and chalky.

Southern Bottletail Squid *Sepiadarium austrinum*

This squid uses two arms to rake sand grains over its head in order to bury itself during the day. (Photos: M Norman) *Below:* When swimming, it uses a large funnel for rapid jet propulsion. (Photo: S Foale)

IDENTIFICATION Pair of kidney-shaped fins on mantle. Colour pattern ranges from large white spots (leucophores) over transparent background to completely bright orange.

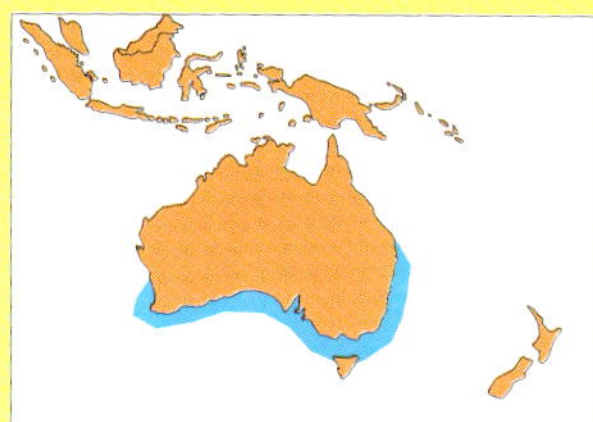

This small squid grows to no more than 4 cm in length. It lives on sandy bottoms in shallow waters and around seagrass beds where it feeds on small crustaceans including isopods. This squid has numerous glands all over the underside of the body. When harassed or captured these glands produce large volumes of slime. The function of this mucus is unknown.

When mating, the pair join head to head and the male passes sperm packages into a pouch under the mouth of the female. The male uses a special ribbed arm to scoop out any previous suitor's sperm. Females lay eggs in bursts of a few at a time, attaching clumps of round eggs at the base of seaweeds or seagrasses.

Lace Bottletail Squid *Sepiadarium sp.*

Right: The Lace Bottletail Squid has kidney shaped fins and lace-like colour pattern. *Below:* In dark colour pattern, showing bumpy mucous glands on underside of mantle. (Photos: M Norman).

IDENTIFICATION Colour pattern ranges from white to deep purple-maroon. Irregular frilled pattern of white on upper surfaces leading to common name.

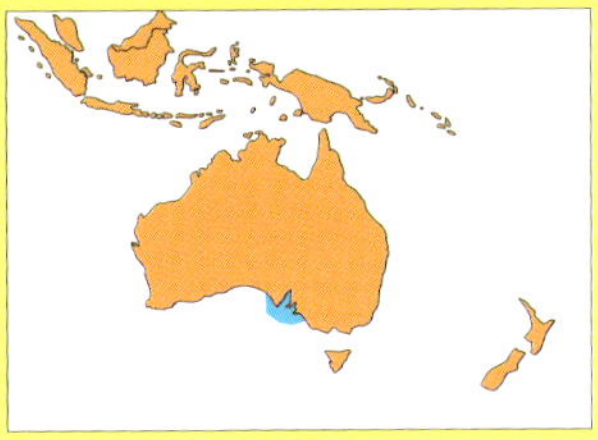

This squid reaches a length of less than 4 cm. It lives on sand, rubble and rocky reef habitats in the shallow gulf waters of South Australia. It remains buried in sand during the day, emerging at night to feed on small crustaceans. It has the same sort of mucous glands all over the underside of the body that are found in the Southern Bottletail Squid and Striped Pyjama Squid. The function of the slime produced when the squid is irritated is unknown.

Little is known of the reproductive biology and behaviour of this species. It has only recently been discovered and awaits a scientific name. Females have a storage pouch for sperm below their beak and males have a modified lower arm similar to that of *Sepiadarium austrinum*.

Striped Pyjama Squid *Sepioloidea lineolata*

These bottletail squid reach about 7 cm long. They are found in sand and rubble, often around seagrass beds to depths of 20 metres. During the day they spend most of their time buried under the sand with only the eyes protruding, a safe way to hide from predators and keep a look-out for passing prey, such as shrimp and fish. The front edges of the mantle opening are fringed with fingers of skin which protrude from the sand when the squid is buried. Their function is unknown.

The round white eggs (about 1 cm long) are attached individually to the underside of rocks or other hard surfaces. The young hatch with a full complement of pyjama stripes, looking like miniature adults. They quickly settle into the sand soon after hatching.

Above: Striped Pyjama Squid, showing raised papillae at front edge of mantle opening. *Inset:* Hiding from predators. (Photos: M Norman)

IDENTIFICATION Distinctive colour pattern of brown to black stripes over white. Eyes sometimes yellow. Paired kidney-shaped fins on mantle. Finger-like papillae on upper side of mantle opening.

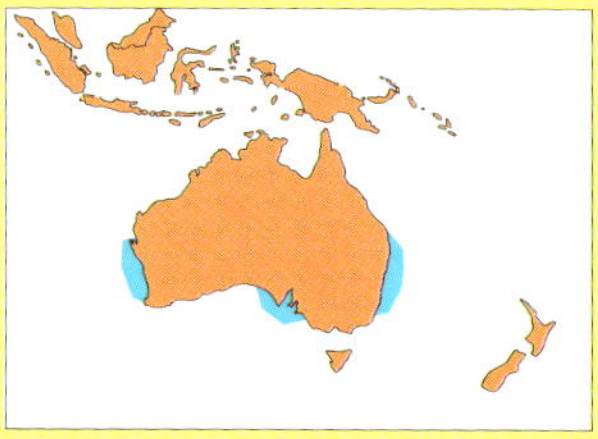

Southern Dumpling Squid *Euprymna tasmanica*

These bobtail squid reach about 7 cm long and are typically found among seagrass beds and soft sediments to depths of at least 80 m. They bury in sand during the day, and if disturbed will emerge wearing a coat of glued sand grains. Acid glands in the skin can quickly shear off this coat for use as a decoy in addition to ink squirting. At night, they use a light organ fuelled by luminescent bacteria to cancel out their silhouette from predators on the sea floor (such as flathead fishes). The males have 2–3 specially enlarged suckers on the second and third arm pairs. Their function is not known.

The eggs of this species are round, and often seen covered in sand grains and attached in clumps around the bases of seagrasses or seaweeds. Young hatch without the bacteria needed for their light organ and must trap bacteria from the water column before the functional light organ can develop.

Above: Resting on seaweed at night, showing brown spots (chromatophores) over an iridescent skin layer. (Photo: M Norman) *Above left:* A sand coat acts as camouflage. (Photo: D Paul)

IDENTIFICATION Large semi-circular fins midway along mantle. Butterfly-shaped light organ in mantle cavity. Four rows of suckers on each arm.

Southern Bobtail Squid *Sepiola sp.*

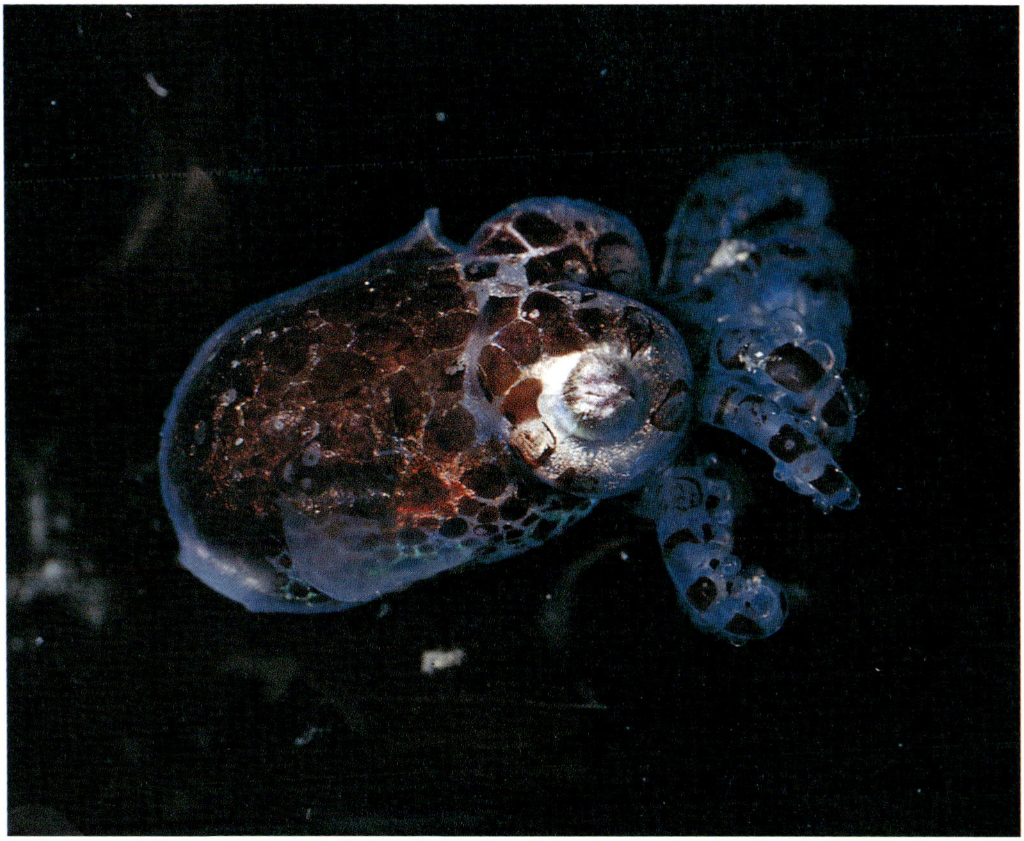

This small squid has not yet been described, nor formally named. Little is known of its biology and distribution. It has been collected at night, actively swimming over rubble, reef and algae in shallow waters. As with the Southern Dumpling Squid, this species harnesses luminescent bacteria in a special light organ within the mantle cavity. Light emitted from this organ is used to conceal the squid from predatory fish. It produces exactly the right amount of down light to cancel out the squid's silhouette when it is swimming in the water column at night, making it invisible from below.

Above: Little is known of the Southern Bobtail Squid. *Below:* Mature males have greatly enlarged suckers on the first and third arms. It is unclear what purpose these suckers serve. (Photos: M Norman)

IDENTIFICATION Grows to about 20 mm long. Small rounded fins on sides of mantle. Body and arms covered in large dark brown spots (chromatophores). Round shiny light organ inside mantle cavity. Two rows of suckers on arms.

Southern Pygmy Squid *Idiosepius notoides*

Above: Colour pattern is typically yellow to brown with a star of white markings around the eye. *Right:* This recently mated female has small sperm packets on the arms and head. (Photos: D Paul)

IDENTIFICATION Distinguished from young of other squid by small round fins at rear end of mantle and cloudy glue-gland on upper surface between fins. Anterior edge of mantle not fused to head. Males have more than 10 suckers on each of the modified lower arms.

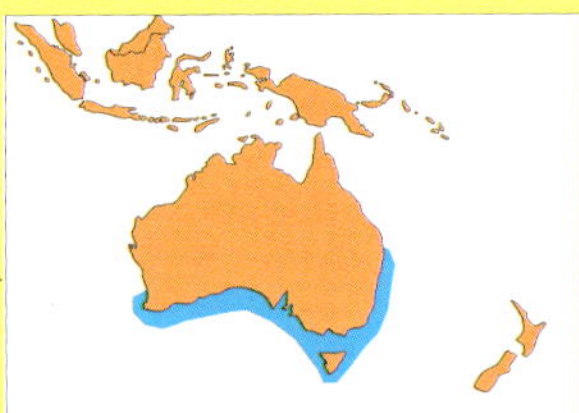

This tiny squid is the smallest of cephalopods, growing to less than 2 cm in length. It lives in seagrass beds, typically eel grass (*Heterozostera* and *Zostera*). It has a special adhesive gland on the rear upper surface of its body which it uses to attach itself to the undersurface of seagrass leaves, where it hides during the day. At night it swims amongst and above the seagrass beds feeding on small shrimp.

These squid approach shrimp from behind, seize the upper side, and rapidly bite through the nerve cord to immobilise them. The female lays single eggs in lines on the underside of seagrass fronds. She shuffles along the frond as she lays the eggs, holding on by the sticky secretion from her adhesive gland.

Tropical Pygmy Squid *Idiosepius pygmaeus*

This species is found in large numbers in shallow coastal areas, particularly among seagrass and mangroves. Reaching less than 3 cm in mantle length when fully grown, the Pygmy Squid has a life span of less than 3 months. Like their southern cousins, these squid use a glue-gland on the end of their mantle to attach to the undersides of algae or seagrasses where they can hide from predators and ambush small crustaceans.

Mature females lay up to 65 individual eggs on a hard substrate, such as rock, then die. The eggs hatch after about two weeks.

Above: Reverse counter-shading. (Photo: A Lewis) *Right:* This day-old hatchling shows tiny rounded fins on the tip of its mantle. (Photo: G Jackson)

IDENTIFICATION Superficially similar to Southern Pygmy Squid. Differs in modified lower arms of male, which have only 13 suckers at the base.

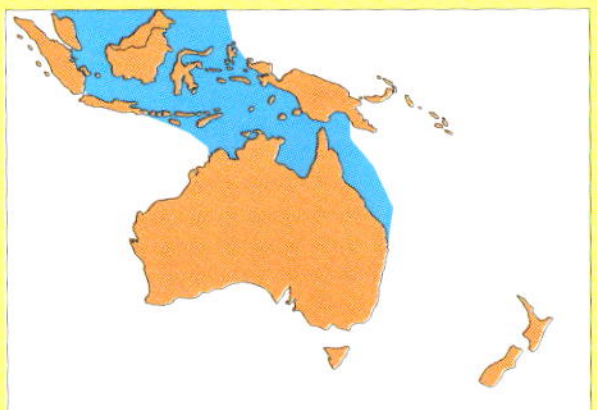

Luminous Bay Squid *Loliolus noctiluca*

Above: Loliolus noctiluca gets its species name from the light organ in its mantle cavity (*noctiluca* = night light). (Photo: Z Florian) *Below:* (Drawing: K Hollis/ABRS)

IDENTIFICATION Squat in shape with large dark spots (chromatophores). Yellow and pink chromatophores on margins of fins and on mantle. Internal shell (gladius) leaf-shaped.

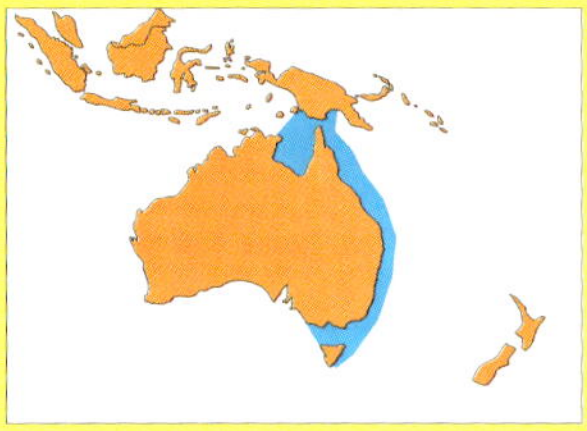

This small squid up to 8 cm in length has large, triangular fins on the posterior end of its mantle. It inhabits shallow coastal areas, including seagrass beds, to depths of about 50 metres. It is sometimes caught during inshore prawn trawls but is not a commercially important species. Within its mantle cavity it has a pair of light organs that are fuelled by luminescent bacteria. The generated light is used to hide the squid's silhouette from below.

Little is known of the biology or behaviour of the Luminous Bay Squid. It appears to have a distribution from the tropics to cool temperate waters, although this may prove to be several closely related species.

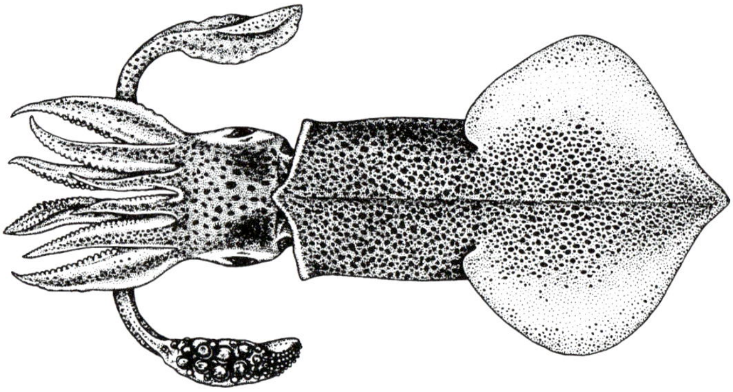

Pencil Squid *Photololigo* spp.

These elongate squid reach lengths of about 30 cm. They form the basis of valuable local and commercial fisheries throughout south-east Asia. Their scientific name comes from the pair of light organs inside the mantle, one on each side of the ink sac. These light organs are powered by luminescent bacteria. The paired, triangular fins can occupy over half the mantle length. As with many other squid, the fins provide stability during the glide after each powerful jet pulse. Molecular studies of this group have shown that there may be many species being treated under single species names.

Above: Pencil Squid form the basis of valuable local and commercial fisheries throughout their range. (Photo: M Norman) *Below:* Fin size and shape varies between different species. (Drawing: K Hollis/ABRS)

IDENTIFICATION Typically covered in red or orange spots (chromatophores). Several species of Pencil Squid found in the region, many identified by the number and shape of teeth on arm sucker rings. Internal shell (gladius) long and thin.

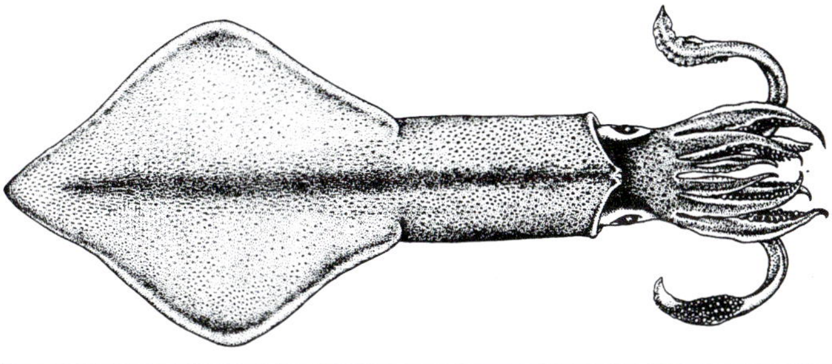

Southern Calamari Squid *Sepioteuthis australis*

Top: The fins of *Sepioteuthis australis* form a diamond shape. (Photo: M Norman)
Above: Transverse bars are frequently shown on the underside while foraging at night. (Photo: M Molloy)

IDENTIFICATION Fins extend along whole length of body and are diamond-shaped, widest at midpoint of body. Colour patterns vary from even orange-brown, to white with black bands, to almost transparent. Internal shell (gladius) quill-shaped.

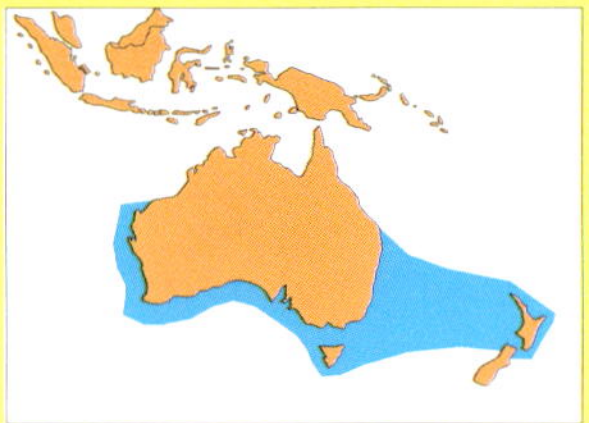

This large squid reaches mantle lengths of up to 50 cm. It is common around southern Australian coastal waters where it typically forms small schools (less than 10 individuals). These schools frequently move over reefs and seagrass beds at night. They are fast, voracious feeders, often approaching divers at night to feed on the fish and crustaceans attracted to the lights. They are popular with anglers, often caught by jig at night off piers and jetties. The black ink squirts of captured squid frequently stain jetties.

The Southern Calamari Squid lays its eggs on the seafloor in fleshy white strings of 2–6 eggs per string.

Bigfin Reef Squid *Sepioteuthis lessoniana*

This medium-sized squid ranges up to about 40 cm in mantle length and is found at depths to about 100 m. It often moves in close to coral reefs at night. It feeds voraciously, mainly on prawns and fish, as well as other crustaceans such as stomatopods and crabs. This species is popular for human consumption and is of importance to major fisheries throughout south-east Asia.

Females lay fleshy strings of 3 to 7 eggs enclosed in finger-shaped capsules. These are attached in clusters to seaweeds, mangrove roots, twigs, stones and under plate corals in shallow coastal waters.

Above: Sepioteuthis lessoniana is identified by the full length fin and transverse bars and spots. *Below:* Two juveniles compete over a recent capture. (Photos: M Norman)

IDENTIFICATION Fins extend along whole length of body, widest towards posterior end to form pear-shape. Iridescent transverse bars or spots on mantle. May be more than one species being treated under this name across tropical Pacific and Indian Oceans. Black bands, to almost transparent. Internal shell (gladius) quill-shaped.

Giant Squid *Architeuthis sp.*

IDENTIFICATION Fins short and heart-shaped at rear of mantle. Key diagnostic character is presence of suckers and knobs along length of tentacle shafts.

Right: A 10 metre female Giant Squid caught off King Island in Bass Strait. (Photo: M Norman) *Below:* Beaks of these animals are commonly found in the stomachs of Sperm Whales. (Photo: D Paul)

Giant squid are the largest of the living cephalopods, with adults weighing more than 250 kg and the length of the mantle exceeding two metres. Total length — including the two long feeding tentacles — has been recorded to 18 metres. *Architeuthis* has long been the subject of folklore, and tales abound of its fights with sperm whales and its ability to drag ships into the ocean depths. In truth, a giant squid is unlikely to be any real match for a sperm whale predator which could be up to 20 times heavier than the squid and probably able to move much faster.

Architeuthis lives on or near the sea floor in depths of 500–1500 m. It feeds on fish and other squid. Suckers and

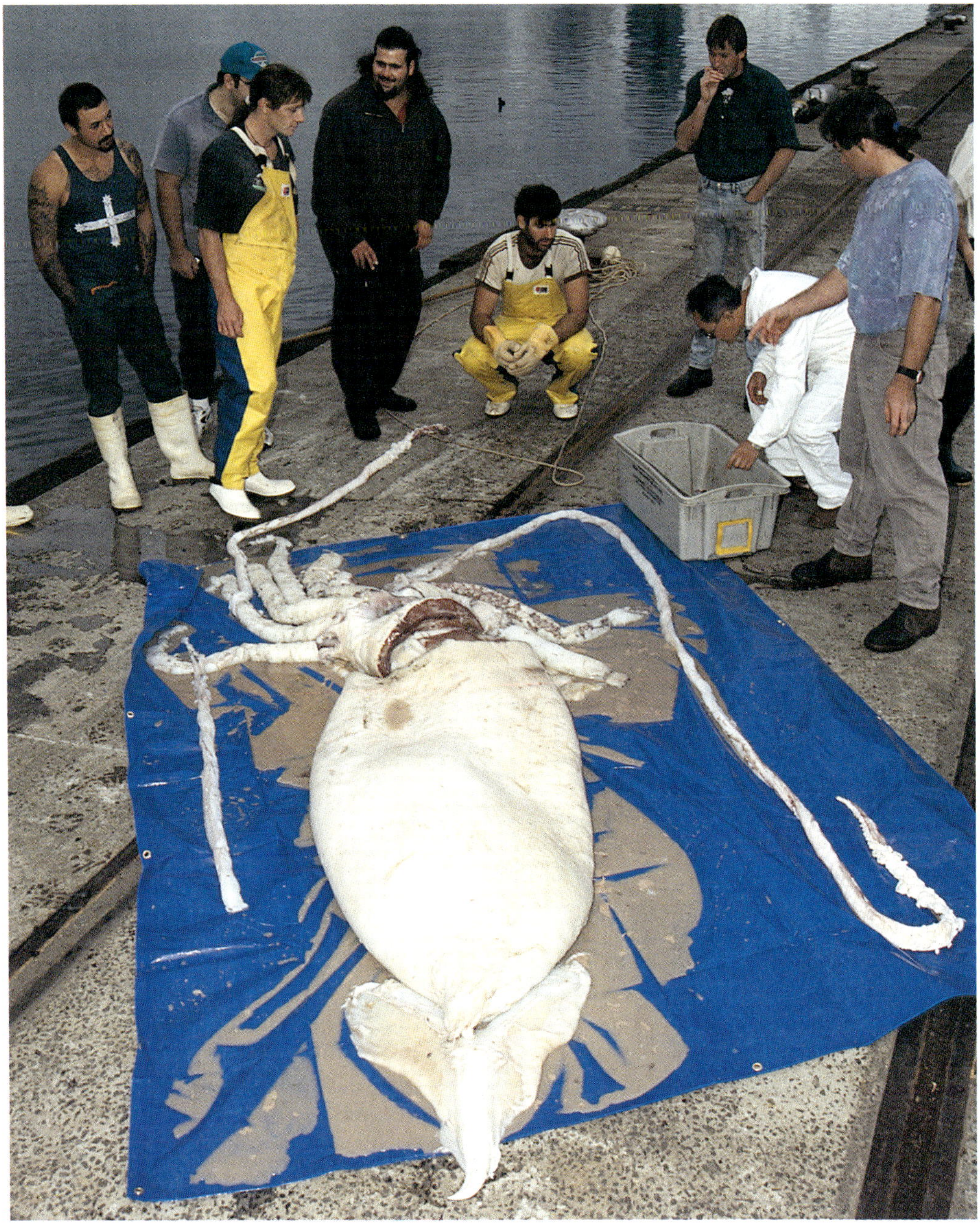

A 12 metre female caught off Tasmania being unloaded in Melbourne. (Photo: D Paul)

knobs along its two long feeding tentacles enable them to be locked together so that their tips act like a snapping claw when grabbing prey. These squid have a high concentration of ammonium ions in the muscles to provide buoyancy (as ammonium is lighter than seawater).

Due to the scarcity of preserved museum specimens in good condition for study, the number of species and their distributions are unknown.

Jewel Squid (various species)

Above: Histioteuthis hoylei from Hawaii showing scattered light organs. (Photo: R Young) *Below:* The left eye of jewel squid is larger than the right. (Drawing: K Hollis/ABRS)

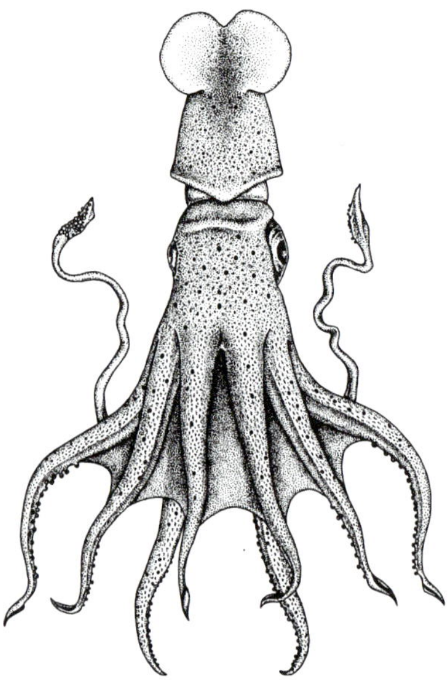

Jewel squid are found in the open ocean at depths of 200–1000 metres. They live in schools, and when small, seem to migrate to shallower water at night, returning to the depths by day. They have numerous light organs (photophores) on the mantle, head and arms. When 'switched on', these photophores serve to mask the animal's silhouette from below through counter-illumination. The photophores may also be used for communication. The body is gelatinous and has a high ammonium content, making them unappealing for human consumption although they are readily preyed upon by fish, dolphins and whales.

The left eye of adult jewel squids is much larger than the right and is pointed upwards in the live animal, presumably searching for the silhouettes of prey swimming overhead.

IDENTIFICATION Many light organs scattered over surface of mantle, head and arms, often against pink or orange background colour. Left eye of adult significantly larger than right eye.

Comb-finned Squid *Ctenopteryx sicula*

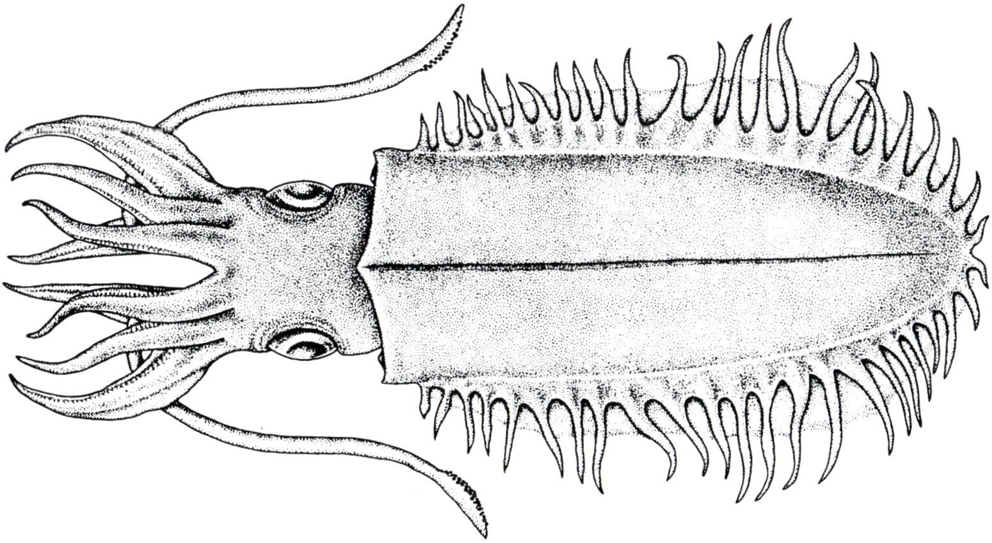

These strange, small squids are less than 10 cm in mantle length. Almost nothing is known of their biology or behaviour. They have light organs (photophores) on the ink sac and within the mantle, presumably to cancel their silhouette. The structure of the fins suggest that they are not capable of sustained, rapid movement. When mature, males develop a large photophore at the posterior end of the body which might be used to attract mates. Little is known about the vertical distribution of these squid, though it seems they inhabit near-surface waters at night.

The light organs of this squid and jewel squid are powered by an enzyme reaction, not the luminescent bacteria used by shallow water species such as the Pencil Squid and the Southern Dumpling Squid.

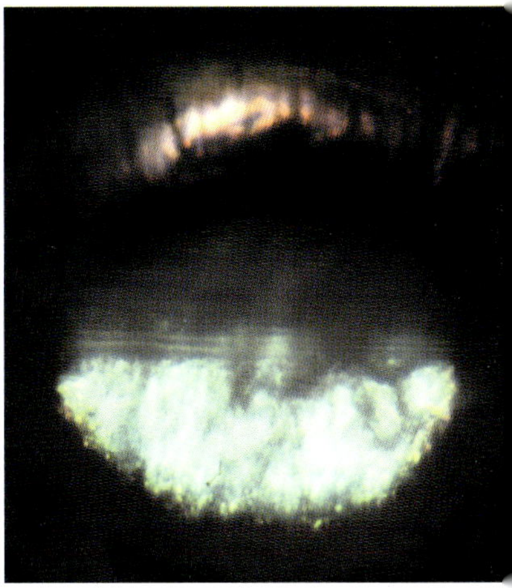

IDENTIFICATION Comb-like ribbed fins extend along length of mantle in adults.

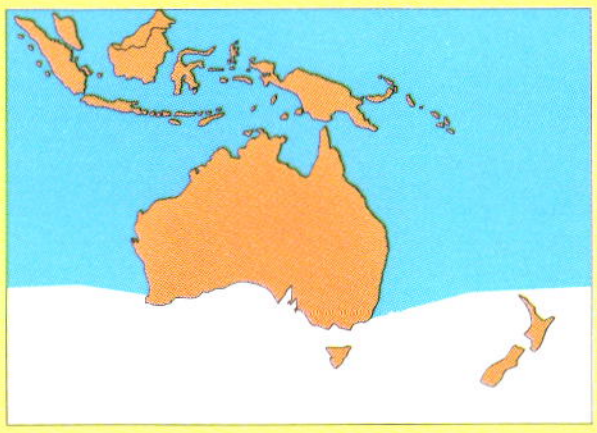

Top: The function of the comb-like fin of this species is not known. (Drawing: K Hollis/ABRS) *Below:* The light organ in the mantle cavity has double reflectors like a car headlight. (Photo: R Young)

Red Arrow Squid *Nototodarus gouldi*

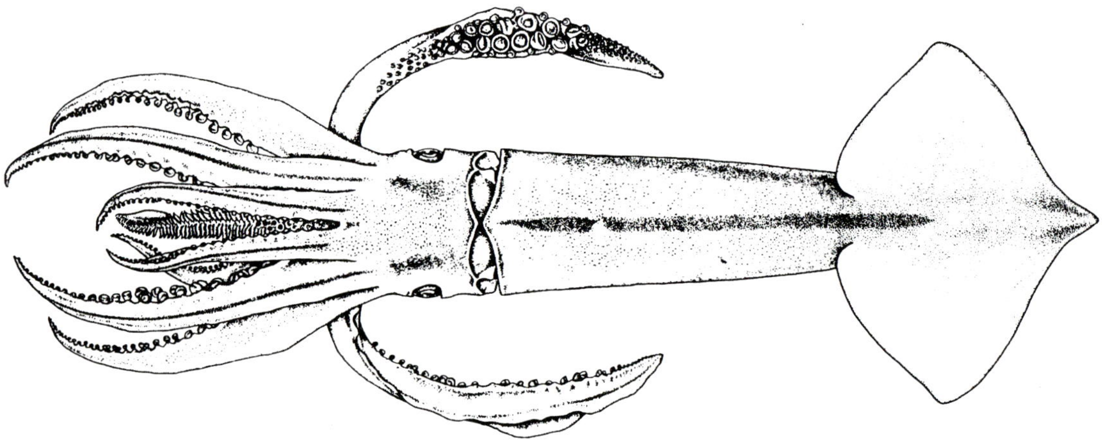

Above: The suckers on the tentacle clubs possess hard rings with large sharp teeth which help grip their prey. (Drawing: K Hollis/ABRS) *Below:* The larval stage (rhynchoteuthion) with a front on view of the feeding tube. (Drawing: K Nolan)

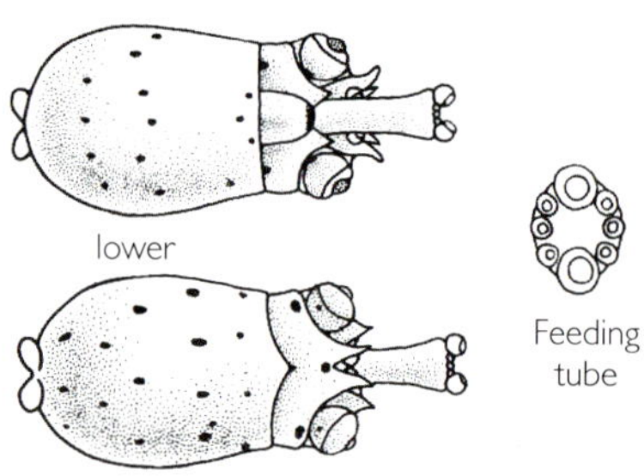

Arrow squid have a muscular, torpedo-shaped mantle up to 40 cm in length. They are found at all depths from the surface to the bottom of the sea, live in large schools and are voracious predators. This squid forms the basis of major jig and trawl fisheries in Australia and New Zealand.

Unlike most squids, whose hatchlings look like miniature adults, squid in this family have a distinct larval stage. This stage is known as the 'rhynchoteuthion' which means the 'nose squid' stage. In the youngest stages (<5 mm mantle length), the two feeding tentacles are fused into a long feeding tube (a 'proboscis') with small suckers around the opening. These larvae and juveniles are found in shallow coastal waters while the adults spend much of their lives off the deeper continental slope and in oceanic waters.

IDENTIFICATION Maroon-red with dark, narrow stripe along midline of upper mantle. Fins are diamond-shaped.

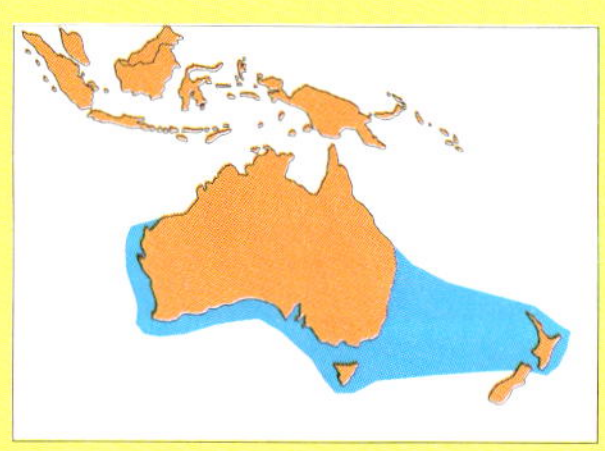

Glass Squid (various species)

These transparent open-ocean squids come in all shapes and sizes, from tiny squids to animals with mantles 1 m long. Their transparency helps hide them from predators. When attacked, some will squirt ink inside their mantles and fill themselves up with water to appear large. Even the silhouettes of the eyes are cancelled out by light emitted from organs on their lower surfaces. The juveniles of many species have huge, bulbous eyes which resume more normal proportions as the squid matures.

Above: Megalocranchia fisheri from Hawaii showing the transparent jelly-like tissue characteristic of these squids. (Photo: R Young) *Below:* The unusual *Bathothauma lyromma* has short arms and stalked eyes. (Drawing: K Hollis/ABRS)

IDENTIFICATION These diverse flimsy squids are transparent when alive. Eyes can be round or stalked in different species. 1–3 crescent-shaped light organs on eyes of many species.

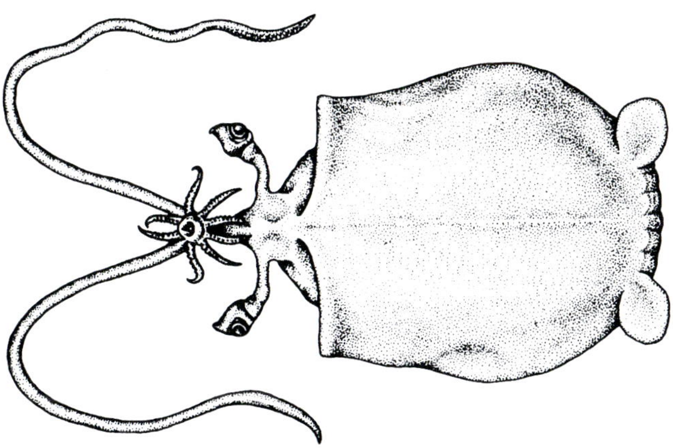

Vampire Squid *Vampyroteuthis infernalis*

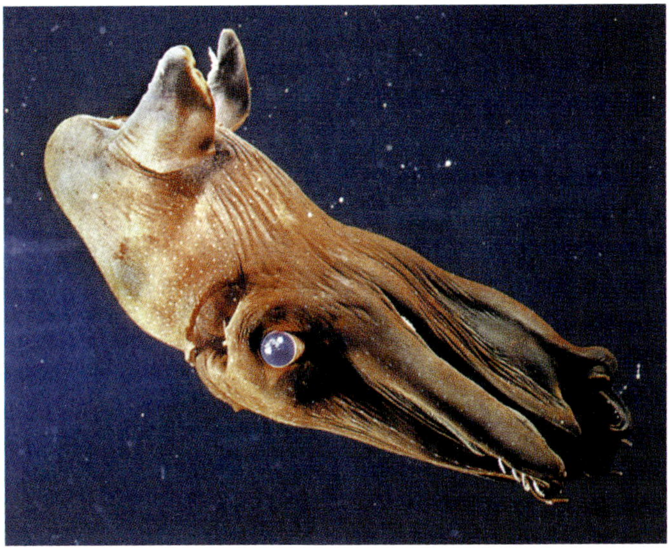

Right: The Vampire Squid swims using the fins on its mantle. *Above:* With its arms and web pulled over its body, it reveals the rows of sensory fingers along each arm. (Photos: K Reisenbichler/MBARI)

This distinctive semi-gelatinous animal lives in moderate to deep waters of open ocean (500–1500 m deep). They reach a total length of around 40 cm. They have eight arms, joined along their length by deep webs, with two long thin filaments emerging from pits between the first and second arms. These filaments may be trailed behind the animal to sense prey. Lines of sensory fingers ('cirri') are present along either side of the single sucker row on each arm.

Vampire squid use medusoid swimming as a means of slow movement, spreading and closing their deeply webbed arms to pulsate through the water. A pair of elongate, paddle-like fins are situated on each side of the mantle of the adult. In the juvenile, there are two pairs of fins.

IDENTIFICATION Maroon to black colour; filaments in pits and large paired light organs on posterior mantle.

Opisthoteuthis spp.

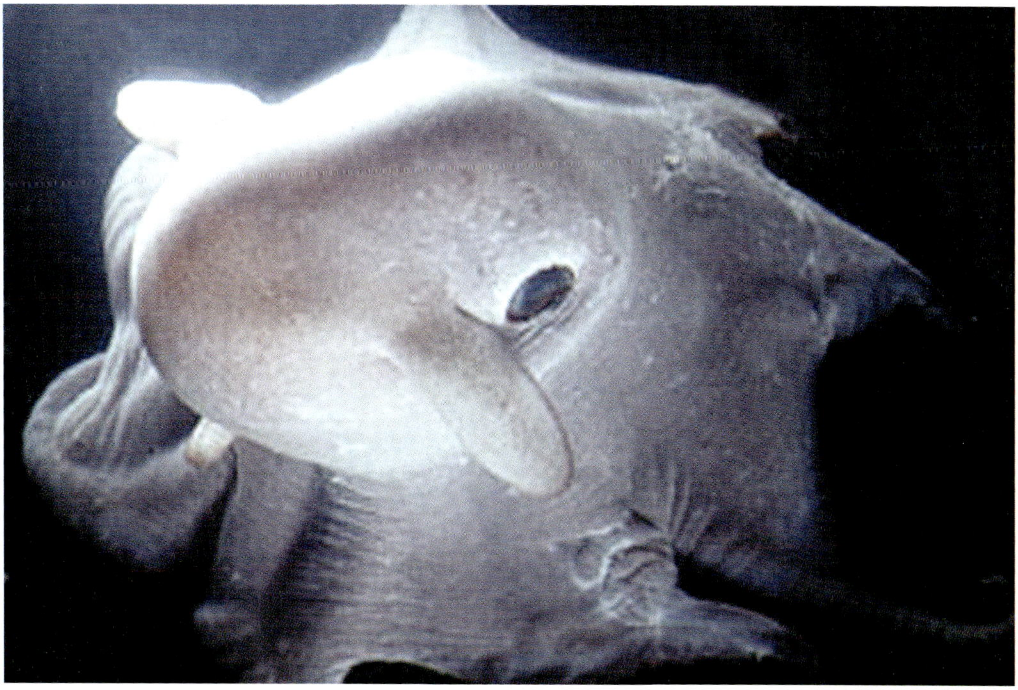

The finned or 'cirrate' octopuses occur in the deep waters of all oceans. These soft gelatinous octopuses can reach a total length of more than 1.5 metres and have been recorded to depths of more than 5 km. Little is known of these octopuses in Australian and Asian waters. Deepsea Finned Octopuses have a pair of fins on the mantle and a single row of suckers along each arm, bordered by rows of sensory fingers ('cirri'). The webs are very deep and these animals often swim with arms and webs spread out, using the fins and jetting to move along. The finger-like cirri may be used to sense the electrical or chemical traces of potential prey.

They are the only octopuses which do not brood their eggs. Eggs are laid in a creased casing and attached to rocks. The males of some species have specially enlarged suckers which may be used in courtship or mating.

Above: Opisthoteuthis grimaldii from the Atlantic. (Photo: M Vecchione) *Below:* Sucker rows and cirri from a specimen from off southern Africa. (Photo: M Norman)

IDENTIFICATION Recognised by fins and rows of fine cirri on either side of suckers. Lack light organs and filaments in pits as are found in Vampire Squid.

Telescope Octopus *Amphitretis pelagicus*

Above: Amphitretis pelagicus has transparent flesh and a single row of suckers. *Below:* It also has tubular telescopic eyes. (Photos: R Young)

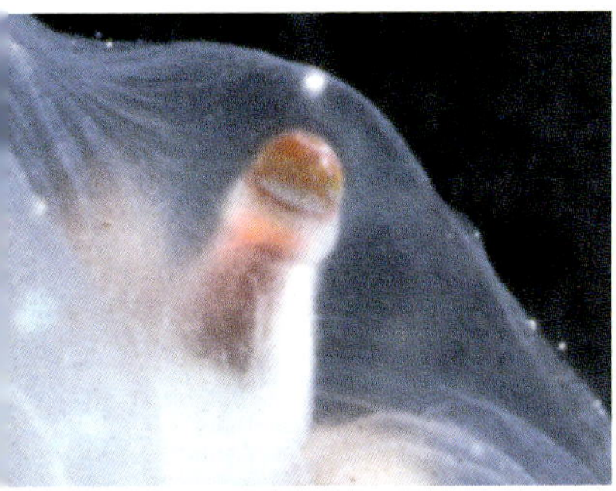

The Telescope Octopus occurs in the open waters of the Indian and Pacific Oceans and the southern Atlantic Ocean. This small free-swimming animal reaches a total length of about 20 cm and is the only octopus in the world with tubular telescopic eyes (left), joined together at the base. These special eyes presumably search the waters above for the silhouettes of their prey, much like the hatchetfishes. This species appears closely related to the Glass Octopus. The underside of the mantle is fused to the funnel, forming two slit-like openings to the mantle cavity.

Nothing is known of life cycle of this species. There is some evidence that smaller animals (less than 30 mm long) occur in shallow waters (<150 m), while larger animals have been collected down to depths of 2000 metres.

IDENTIFICATION Easily identified by telescopic tubular eyes and two slit-like openings to mantle cavity. Single row of suckers on arms.

Glass Octopus *Vitreledonella richardi*

The Glass Octopus is a strange free-swimming octopus that lives in the open waters of tropical and subtropical oceans. It reaches a total length of about 45 cm and has been caught in trawl nets in all oceans of the world from surface waters to over two kilometres deep. It spends its life in midwater, never touching the sea floor. Its transparent form probably makes it harder for predators to see. Even its brain is small and hard to see (right).

Little is known of reproduction in this octopus. The male has single enlarged suckers on the arms beyond the web margins and the female is thought to brood her eggs in the space formed when her arms are held together.

Above: Vitreledonella richardi has a cigar-shaped digestive gland (liver) which it keeps vertical to reduce its silhouette from below. *Below:* The optic lobes of its brain are stalked, and situated close to the eyes. (Photos: R Young)

IDENTIFICATION Transparent form, rectangle-shaped eyes and single wide opening to mantle cavity. Single row of suckers on arms.

Banded String-arm Octopus *Ameloctopus litoralis*

Above: Ameloctopus litoralis feeds by poking its long arms down crevices and under rubble to catch small crustaceans and fish. *Below:* This individual has discarded one of its arms. (Photos: M Norman)

This small species has the longest, thinnest arms (relative to its body size) of any Australian octopus. Its arm span can exceed 30 cm. It lives on intertidal mud and rubble flats, emerging at night low tides to feed. It is able to drop its arms as a decoy to predators, the discarded arms continuing to writhe for up to 5 hours. Meanwhile, the animal crawls away and grows new arms. This octopus is one of the few shallow-water species which lacks an ink sac. Perhaps its arms are too long to be hidden by an ink cloud.

The males of most other octopuses have, from birth, a specially modified shorter arm, which they carefully guard, and which is used to pass sperm packages to females. In *Ameloctopus*, young males have full-length arms, with none modified for transferring sperm. When they reach maturity, one arm drops off, and in its place the specially modified arm grows from the stump.

IDENTIFICATION Recognised by long banded arms, small body and tiny eyes.

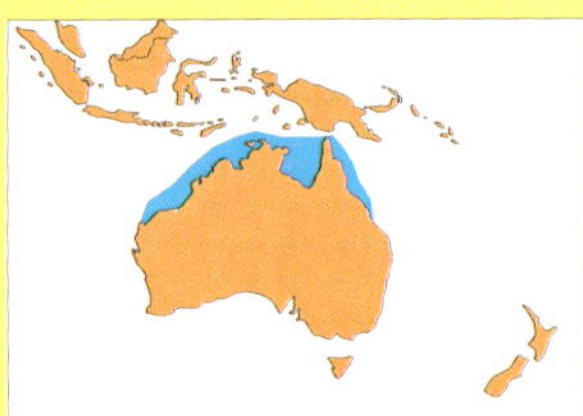

Velvet Octopus *Grimpella thaumastocheir*

This robust octopus has an armspan which reaches about 60 cm. It occurs on rocky reefs and areas of rubble and sand and is only known from southern Australia. Its most dramatic feature is the absence of an ink sac, unusual for a shallow-water octopus. Many deep-water species lack an ink sac, ink decoys being useless in dark depths.

Little is known of the biology of this species. Mature males possess specially enlarged suckers which may be used for courtship displays or other mating behaviour. This species lays large eggs which means that the hatchlings would be well developed and settle to the seafloor soon after hatching.

Above: The green iridescent sheen of *Grimpella thaumastocheir* gives it a velvet appearance and hence its common name. *Below:* A female guards her eggs, laid on the underside of a rock. (Photos: M Norman)

IDENTIFICATION Pink to blood-maroon in colour with green iridescent sheen. Skin sculptured with low irregular bumps.

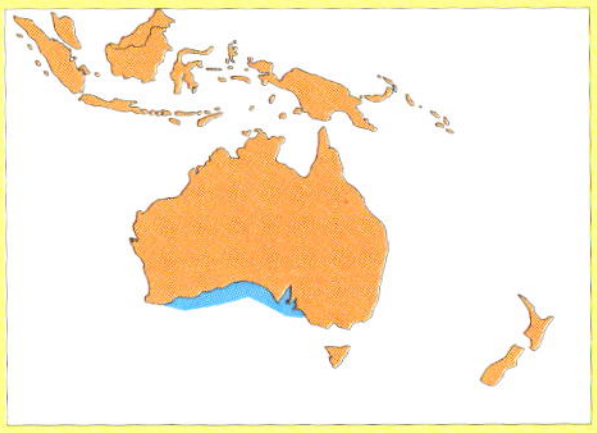

Blue-lined Octopus *Hapalochlaena fasciata*

Top: The Blue-lined Octopus in full alarm display, advertising its poisonous nature. *Above:* When resting or stalking prey, it has excellent camouflage. (Photos: M Norman)

NOTE In 1967 a 23 year-old Sydney man was bitten while examining a Blue-lined Octopus on the back of his hand. A tiny pair of bruises were found on one of his knuckles where the octopus had injected a paralysing tetrodotoxin. He was unable to breathe and died by asphyxiation.

This small octopus has an armspan of less than 20 cm. It lives on rocky reefs and around seagrass beds in shallow coastal waters. It is often found in bottles or empty gastropod and bivalve shells. It feeds on crustaceans and fish, using a strong toxin to paralyse large prey.

The males of many octopus species sit at some distance from the female when mating, presenting an arm to her along which sperm packages are shunted. This octopus mates while the male is carried around on the mantle of the female. The female Blue-lined octopus plaits her large eggs into strings and carries them in her web until they hatch.

IDENTIFICATION Drab light brown with fixed white spots when not alarmed. Blue lines (not rings) on mantle.

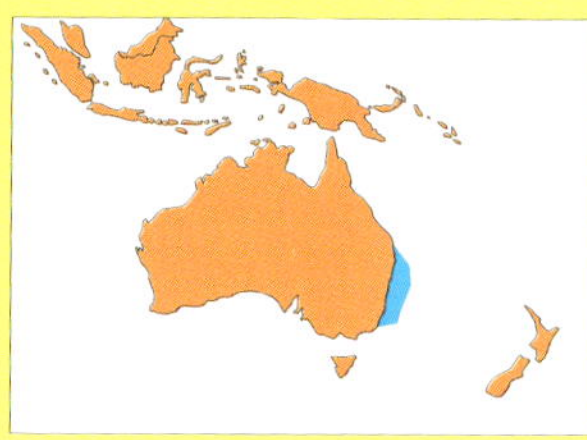

Greater Blue-ringed Octopus *Hapalochlaena lunulata*

This dramatic small octopus occurs in shallow tropical waters, where it is often found foraging on intertidal reef flats at night. It usually reaches an armspan of less than 15 cm. It feeds on fish and crustaceans which it paralyses by injecting saliva containing the powerful toxin, tetrodotoxin. More than one large-ringed species may be being treated under this name throughout the tropical western Pacific Ocean.

As with other blue-ringed octopuses, the female of this species does not attach eggs to rocks, coral or other solid surfaces, instead she carries a string of eggs with her, held in the trailing web and arms.

Above: (Photo: C Bryce) *Below:* The elongated mantle of the Greater Blue-ringed Octopus is often referred to as a 'pixie cap'. (Photo: M Norman).

NOTE The Greater Blue-ringed Octopus has been responsible for at least one human death, in Darwin in the 1950s.

IDENTIFICATION Large obvious rings on mantle and arms. Mantle is often extended into an elongate tip.

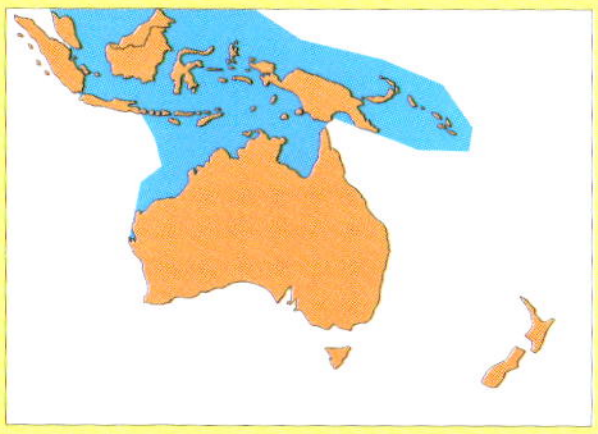

Southern Blue-ringed Octopus *Hapalochlaena maculosa*

Above: A female Southern Blue-ringed Octopus carrying eggs. (Photo: M Norman) *Below:* Day-old hatchlings quickly settle to the seafloor and commence foraging. (Photo: D Paul)

This small octopus, with an armspan typically less than 15 cm, lives on rocky reefs and around seagrass beds. It feeds on crabs, fish, shrimp and shellfish, using a strong toxin to paralyse large prey. This toxin is produced by bacteria in the salivary gland and is injected when the octopus bites with its beak. Although not directly linked to any human fatalities, near-deaths indicate the strength of the toxin produced by this octopus. Despite their venomous nature, blue-ringed octopuses are not aggressive and are more keen to escape than attack humans.

As in other blue-ringed octopuses, the females do not attach eggs to a lair or shells, instead carrying the eggs in the web. The hatchlings already possess small blue rings and are thought to be toxic from birth.

IDENTIFICATION Drab light brown when not alarmed. When disturbed or harassed it will flash small (1–2 mm) iridescent blue rings on mantle and arms.

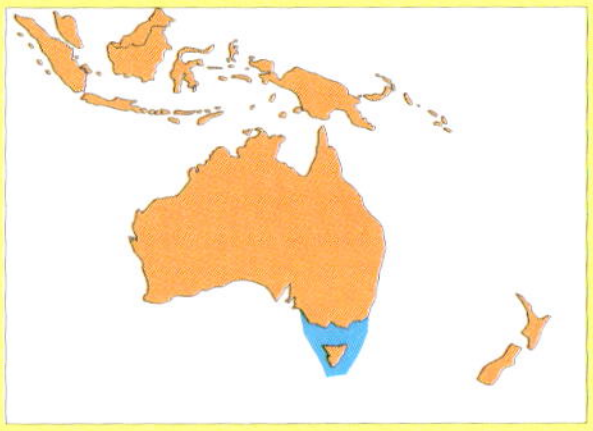

Marbled Octopus *Octopus aegina*

This small octopus has an armspan of about 40 cm. It lives on soft sediment substrates from India to the Philippines. It is commercially harvested in many areas and forms the basis of a major trawl fishery in the Gulf of Thailand. Along with a related species with ocellate (false) eye-spots, the Marbled Octopus is exported frozen from Thailand to markets all around the world, often marketed under the name 'baby octopus'. It frequently lays eggs in shells and discarded bottles.

The sperm packages of this species are armed with sharp teeth. As the sperm package unfurls in the female oviduct, these teeth grip the oviduct walls and work the sperm package all the way into the ovary.

Above: Octopus aegina, showing its marbled colour pattern, consuming a fish. *Below:* Showing the pale stripe along the centre of its mantle. (Photos: M Norman)

IDENTIFICATION Dark reticulations form a marbled effect over arms and mantle. Pale longitudinal stripe along upper midline of mantle. Narrow pale bar across head in some colour patterns.

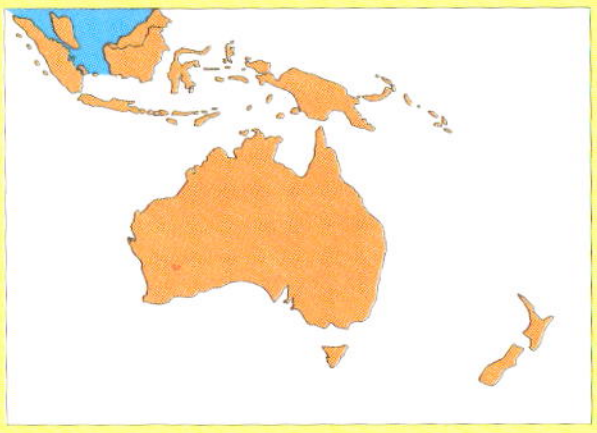

Capricorn Octopus — *Octopus alpheus*

Above: The Capricorn Octopus has distinctive white spots over mantle and arms. *Below:* When hunting, it uses its longer front arms to feel down crevices and burrows. (Photos: M Norman)

This octopus reaches an armspan of about 40 cm and is only known from a single group of islands on the Great Barrier Reef. It emerges during night low tides to forage in the shallow pools of exposed intertidal reefs, feeding primarily on crabs which it paralyses and takes underground to eat while the tide is in.

It will crawl from pool to pool over dry coral and rock and can sometimes be found by the squeaking noises made when it squirts air through its funnel, instead of water. As the tide returns, it quickly buries in sand or rubble.

This species lays large eggs which hatch into well-developed young capable of quickly adopting a bottom-dwelling life. The limited ability of these young to disperse may account for the restricted distribution of this species.

IDENTIFICATION Large white spots over red base colour. Front arms longer and thicker than other arms.

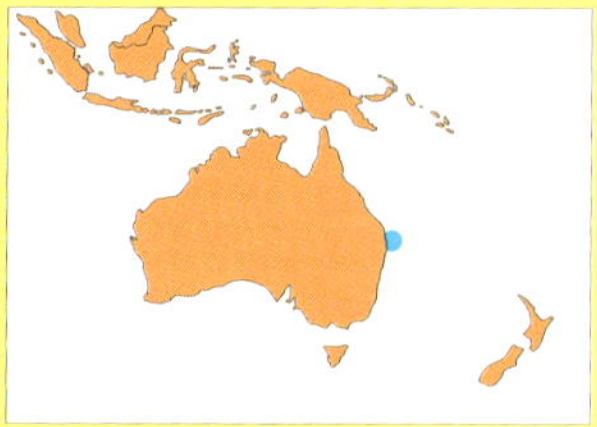

Hammer Octopus *Octopus australis*

This octopus has an armspan of about 40 cm. It occurs primarily on sand habitats and around seagrass beds from shallow waters to depths of about 100 m. It buries in the sand during the day, emerging at night to feed. It may also use discarded human refuse such as cans or bottles as an artificial shelter. The edge of its body has a skin ridge or 'keel' which may help to guide it into the sand when burying itself. This species is commercially harvested as a trawl bycatch in Queensland and northern New South Wales.

The common name refers to the large, club-like modified arm tip found in mature males. The primary function of this structure is to pass sperm packets to the female.

Above: Octopus australis has a keel-like ridge on the sides of its mantle. *Below:* The hammer-like modified arm-tip of a male Hammer Octopus. (Photos: D Paul)

IDENTIFICATION Typically sandy cream in colour with keel-like ridge along edge of mantle. Males with large hammer-like modified tip on third right arm.

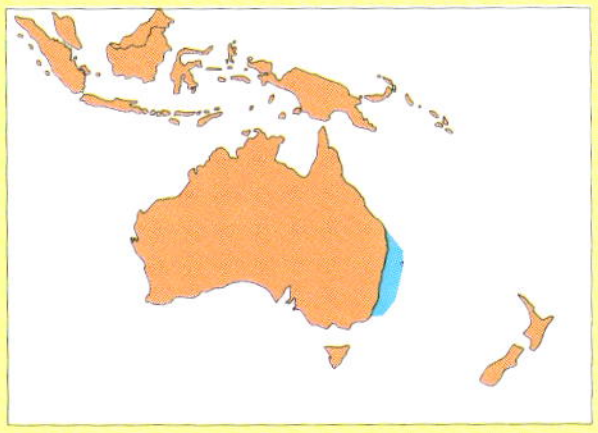

Southern Keeled Octopus *Octopus berrima*

Above: Octopus berrima moving over sand and raising its coiled arms to impersonate drifting seaweed. (Photo: M Molloy) *Below:* It can hide in sand and raise one eye like a periscope to see if the coast is clear. (Photo: S Foale)

This octopus has an armspan of about 50 cm. It is common on sand habitats and around seagrass beds in shallow waters, but has also been captured down to depths of more than 250 metres. It forms the basis of small-scale octopus pot fisheries in South Australia and Victoria. Like its relative, the Hammer Octopus, it hides in the sand during the day or uses rocks or human rubbish as shelter; it also has the keel-like ridge around its body. If approached under-water, it will often bury quickly in the sand. It feeds primarily on crabs.

The female Southern Keeled Octopus lays large eggs which are attached singly to rocks or inside shells or bottles. The hatchlings are well-developed and immediately settle on the seafloor and start foraging.

IDENTIFICATION Typically cream to light brown. Skin keel around mantle edge. When agitated it flares webs and darkens around eyes.

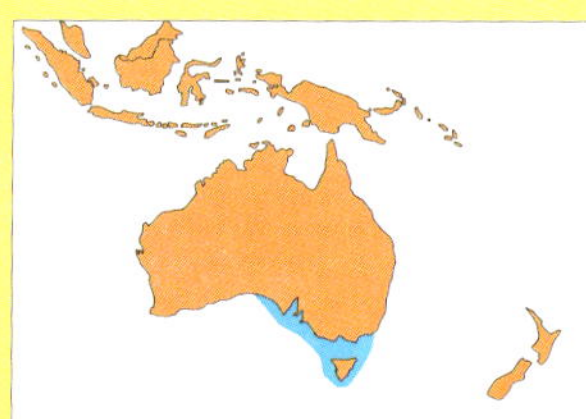

Southern White-spot Octopus *Octopus bunurong*

This octopus lives on sand substrates where it emerges at night to forage, probably hiding in the sand during daylight hours. Its diet is unknown. Like the Southern Sand Octopus (*Octopus kaurna*), it can rapidly disappear under the sand. It has a short skin ridge along the sides of the mantle, known as a lateral ridge. This ridge may act as a keel, helping to guide the mantle under the sand when burying itself.

This species lays large eggs which hatch into well-developed young, which would quickly settle on the seafloor after hatching. Nothing else is known of its reproductive behaviour.

Top: Little is known of *Octopus bunurong* as it was only recently discovered. *Above:* It has skin ridge along each side of its mantle. (Photos: D Paul)

IDENTIFICATION Small white spots over red mantle with regular transverse bands of small spots along arms. White skin ridges on each side of mantle only extend for 75% mantle length.

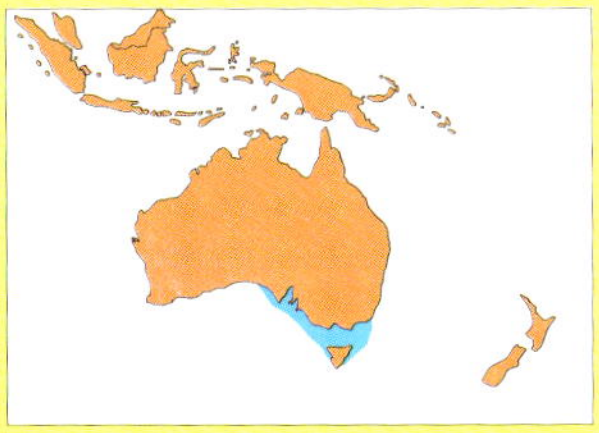

Day Octopus — *Octopus cyanea*

Above: The false eye-spots of *Octopus cyanea* give the impression of a much larger animal. *Below:* It is capable of raising large fingers of skin over its body and arms to impersonate corals and other backgrounds. (Photos: M Norman)

This large muscular octopus has an armspan that reaches over 2 metres. It can weigh more than 6 kg. It is commonly seen on coral reefs across northern Australia and throughout the Indo-West Pacific region. It is active mainly during the day and is a camouflage expert, moving over diverse habitats and rapidly modifying both its colour and sculpture to perfectly match backgrounds. This octopus mainly feeds on large crabs but will also take fish. Its lair is easily recognised by the scatter of large clean crab carapaces (shells) surrounding a hole.

This species lays up to 600 000 eggs only 1–2 mm long. The tiny young enter the plankton and spend weeks being transported in ocean currents from reef to reef. Sexual cannibalism has been observed in this species; large females eat the male after mating.

IDENTIFICATION False eye-spot on each side of arm crown, between bases of arms 2 and 3. Regular rows of white spots over maroon arm tips.

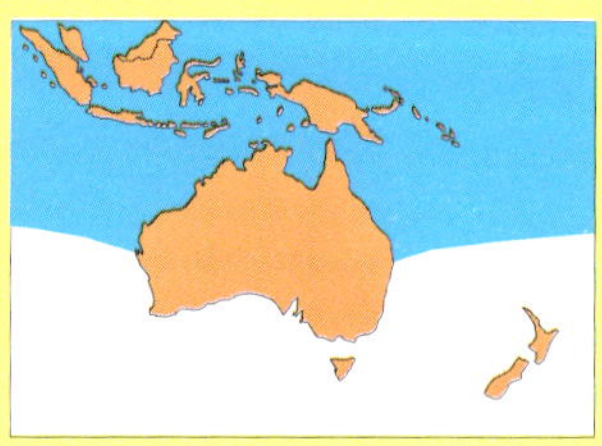

Red-spot Night Octopus *Octopus dierythraeus*

This large muscular octopus has an armspan which reaches at least 1.5 metres. It lives in coastal tropical waters on mud, sand and rubble substrates. It is commonly seen at night during low tides on intertidal flats. This species occupies semi-permanent lairs excavated under coral heads or rubble which can easily be recognised by the accumulation of intact shell remains, particularly bivalve shells, around the mouth. It opens larger shells by pulling apart or drilling and poisoning the shellfish using its toothed tongue (radula).

This species produces large eggs, indicating that the young would be well-developed and bottom-dwelling soon after hatching. No brooding females have ever been found.

IDENTIFICATION Typically mottled orange-red with white spots over body and arms. Front arm pair longest and most robust. When disturbed it displays red spots on a white background, the inverse colour pattern of many related species.

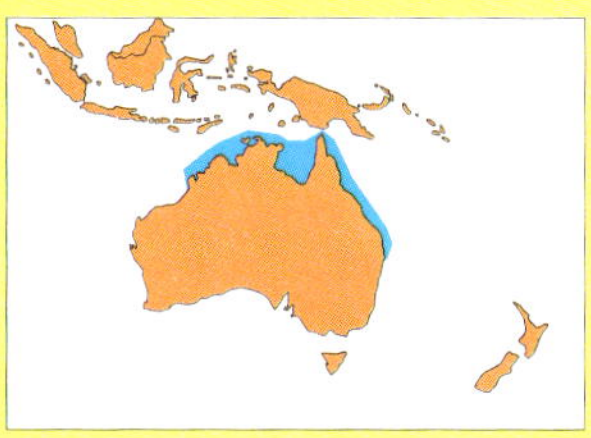

Above: Octopus dierythraeus in resting colouration. *Below:* When disturbed or threatened, it transforms to the inverse colour pattern of red spots over a white body and arms. (Photos: M Norman)

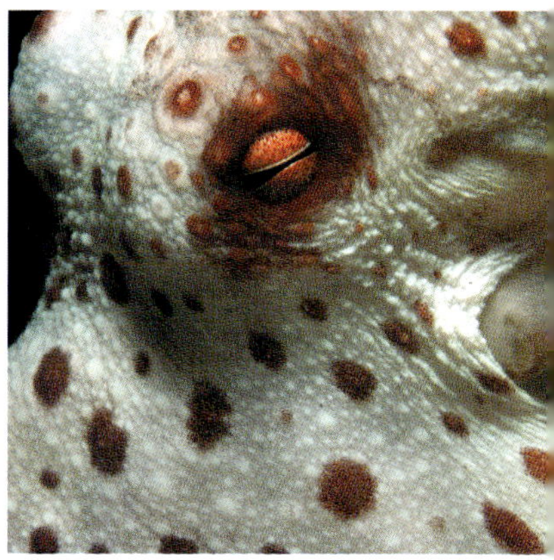

Southern Sand Octopus *Octopus kaurna*

Above: Octopus kaurna often walks around at night with its body drawn into a long cigar shape. (Photo: M Norman) *Below:* Mating swarms have been observed a number of times on shallow sand areas. (Photo: L Reynolds)

This elongate octopus has an armspan which reaches 50 cm. It lives predominantly on sand flats where it forages at night. It is an excellent burier and can quickly draw itself below the surface, either by threading its fine arms into the sand and dragging the body down, or loosening the sand with fast jets of water and diving in. It feeds primarily on crustaceans. During the day this octopus excavates a temporary cavity below the sand, the sand grains being bound together by mucus from the arms.

Little is known of the reproductive biology of this octopus, other than it produces large eggs so that the hatchlings would be well-developed and crawl away like miniature adults. No brooding females have ever been found.

IDENTIFICATION Typically plain to cream over body and arms. Lacks numerous white spots of related night species. When elongated, it often shows a longitudinal maroon band along each side of the mantle which extends down the arms.

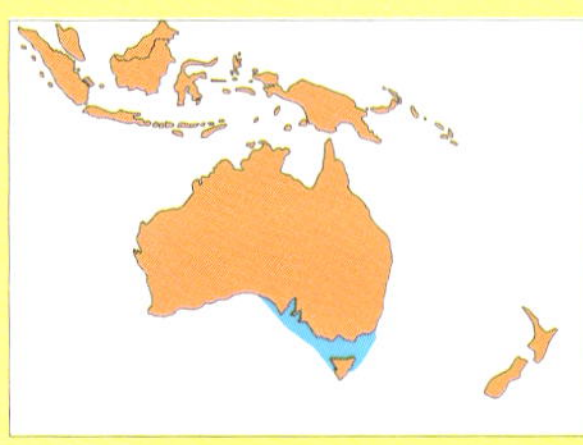

Maori Octopus *Octopus maorum*

This octopus reaches an armspan of over 3 metres and a weight of more than 10 kg. It occurs on or near reefs where it excavates large lairs under boulders. It feeds on crustaceans, fish and shellfish, including abalone. It has been known to take rock lobsters out of lobster pots. It mainly feeds at night, but is sometimes observed in intertidal rock pools foraging during the day for large crabs. It forms the basis of a commercial fishery in southern Tasmania where some of the catch is pickled in the Greek style.

This species lays numerous small eggs. The tiny young (2–3 mm long) enter the plankton and are carried in surface currents across the Tasman Strait to New Zealand.

IDENTIFICATION Robust and muscular. Orange-brown to purple-grey with numerous small white spots on the upper surface of arm bases. Skin on mantle can be raised as large spikes over upper surfaces.

Above: Octopus maorum is the largest octopus species found in the region. *Below:* It can raise spikes of skin over its body which act as camouflage when foraging through seaweeds. (Photos: M Norman)

Veined Octopus *Octopus marginatus*

Above: Octopus marginatus can display blue suckers and a light triangle under each eye. (Photo: C Griffith) *Below:* It feeds by probing with its arms down holes in the mud to catch shrimp and fish. (Photo: R Steene)

This octopus reaches an armspan of about 50 cm. It is found mainly on sand and muddy/shelly habitats from shallow waters to 190 metres. Little is known of its biology and behaviour. It occurs across all of the tropical Indian Ocean from Durban (South Africa) to Australia, New Guinea and southern Japan. It is harvested in south-east Asia by trawl and traps where it is known as 'saa liu' (sand bird).

This species lays up to 100 000 small eggs which hatch into tiny planktonic young. These young are carried in ocean currents, which account for its wide distribution.

IDENTIFICATION Branching veins of dark brown over body and down arms. Edges of arms often darkened in contrast with white to blue suckers. White to cream wedge under each eye.

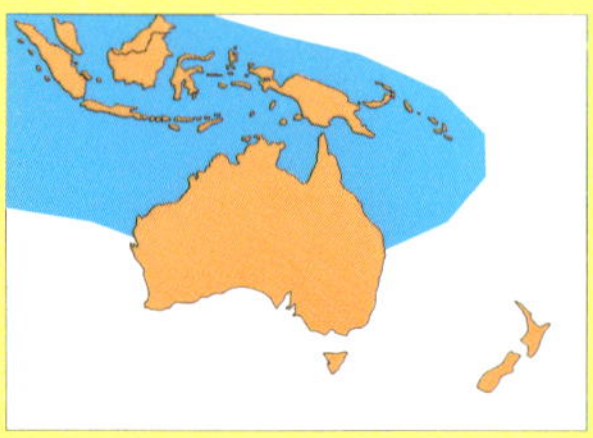

Poison Ocellate Octopus *Octopus mototi*

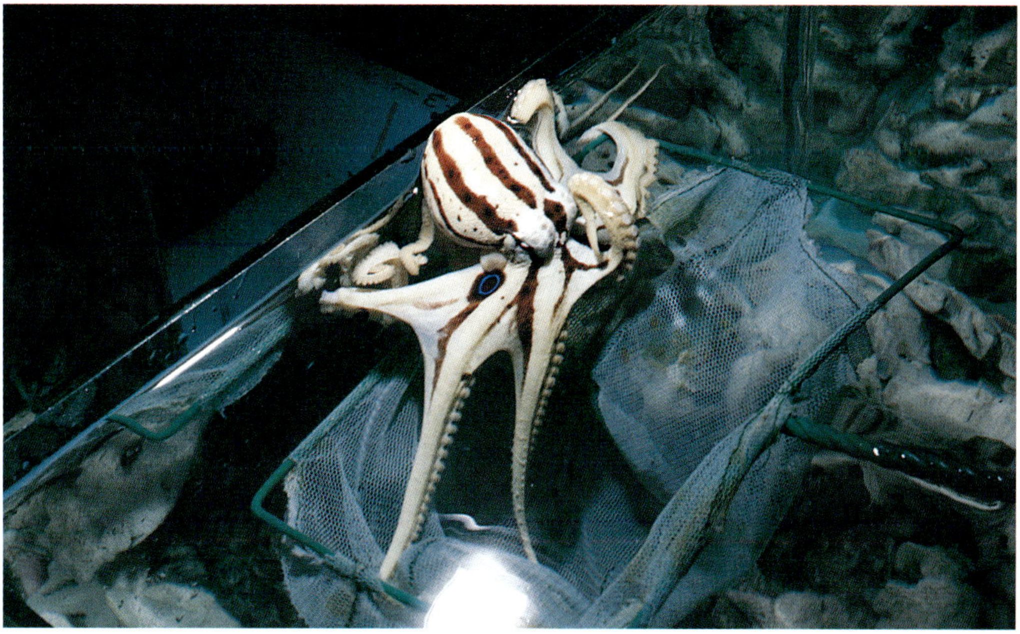

This octopus reaches an armspan of about 50 cm. It occurs on coral reefs, coral rubble and sand habitats from shallow waters to at least 50 metres. It appears to feed on hermit crabs as its lair is often surrounded by numerous coiled shells which have been drilled twice; once by a predatory snail who killed and ate the original shellfish, and the second time by the octopus in order to paralyse and extract a hermit crab occupant. The common and species names derive from Rapa Island in the South Pacific where the locals know it as *fe'e mototi* (poison octopus). The distinctive colour pattern and iridescent blue rings may be advertising a venomous nature.

The females of this species lay fairly small eggs in long branching strings (festoons). The young are likely to be planktonic, accounting for the wide distribution from Okinawa (south of Japan) to Australia and the South Pacific.

Above: Octopus mototi in alarm display showing an iridescent blue ring within its false eye-spot. *Below:* When resting, its normal colour pattern is orange-brown. (Photos: M Norman)

IDENTIFICATION Normally drab orange-brown. When alarmed, suddenly changes to white background with maroon stripes along body and arms. Black 'petal'-like spots over each eye. Pair of iridescent blue rings act as false-eye spots to scare off potential predators.

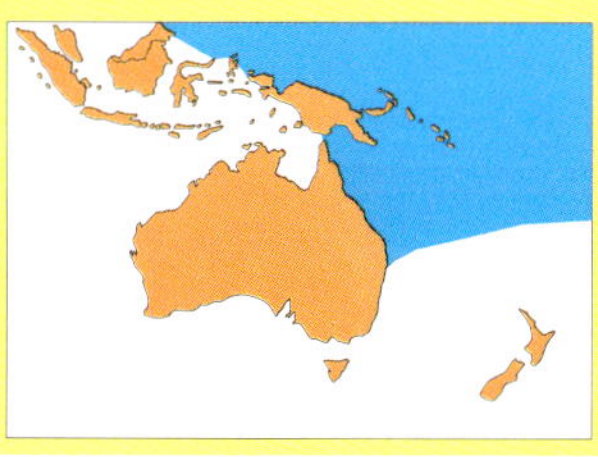

White-striped Octopus *Octopus ornatus*

Above: Octopus ornatus has regular white spots along its sucker margins. (Photo: M Norman) *Below:* It can raise long white flaps of skin within its mantle stripes. (Photo: R Fenwick)

IDENTIFICATION Upper arms much longer and thicker than other arms. Pink to red in colour with paired white spots down arms and short longitudinal white stripes on mantle.

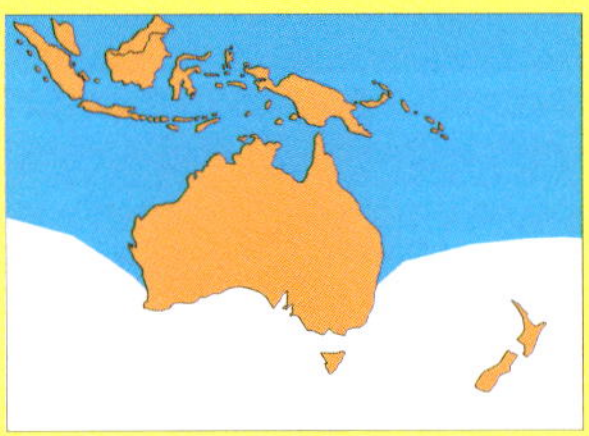

This long-armed octopus has an armspan of up to 2 metres and can weigh several kilograms. It is active at night on coral reefs throughout the tropical Indo-Pacific region. It spends the day buried under sand or rubble, closing the entrance to these temporary lairs with pieces of coral rubble. It has a diverse diet but has a particular taste for other octopus species. The stomachs of these octopuses are full of the beak remains of other octopuses.

This species lays tens of thousands of small eggs, each about 2 mm long. The tiny hatchlings enter the plankton and are carried in surface currents over areas of deep water between their coral reef habitats.

Pale Octopus *Octopus pallidus*

This robust octopus reaches an armspan of 60 cm. It occurs on sand substrates, often in association with sponge gardens or beds of large solitary sea squirts. It emerges at night to feed on crustaceans and shellfish, hiding during the day in rubble, empty sea squirt cases or human rubbish. In the past, this species has been harvested in Port Phillip Bay. It has been recorded from shallow waters to at least 275 metres. It is often found in the same habitat as the Southern Keeled Octopus.

This octopus lays large eggs which are attached singly to the roof of crevices or discarded bottles. The well-developed hatchlings already have good skin sculpture and camouflage. They crawl away after hatching, and immediately start to forage.

Above: Octopus pallidus in a spiked posture showing branched fingers of skin. *Below:* Showing regular oval patches on its skin. (Photos: M Molloy)

IDENTIFICATION Robust muscular species with numerous regular pavement-like patches over entire body and arms. Pointed and branched digits of skin can be raised on its body. Lacks skin ridge (keel) of Southern Keeled Octopus.

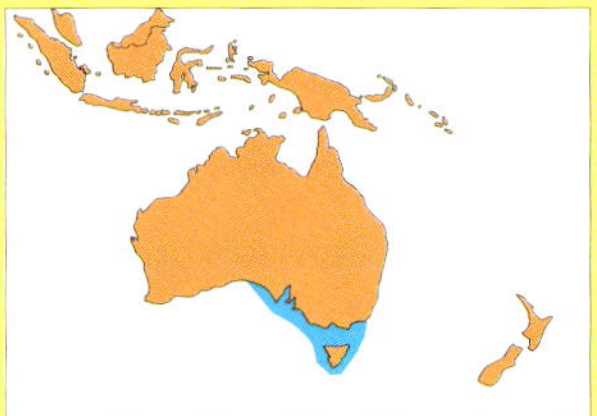

Frilled Pygmy Octopus *Octopus superciliosus*

Above: Octopus superciliosus gets its common name from the frill of white flaps around the edge of its mantle. (Photo: M Molloy) *Below:* It can raise three large fingers of skin over its mantle and one over each eye. (Photo: M Norman)

This small octopus has an armspan of only 10–15 cm. Virtually nothing is known of its biology as it has rarely been encountered and observed in the wild. The only known observation of the natural behaviour in this species was an animal found fully camouflaged, hitching a ride on the back of a large shellfish (a wavy volute). Several live animals have been captured by dredges on shallow seagrass beds. As with other pygmy species, they may spend most of their lives hidden amongst the roots of seagrasses and seaweeds or deep within rocky crevices, foraging for small crustaceans and fish.

This species lays large eggs, so the young are likely to crawl away on hatching, not entering the plankton as occurs in small-egg species.

IDENTIFICATION Narrow arms and flap-like white frill around edge of mantle. Background colour ranges from cream, to orange, to dark maroon-brown.

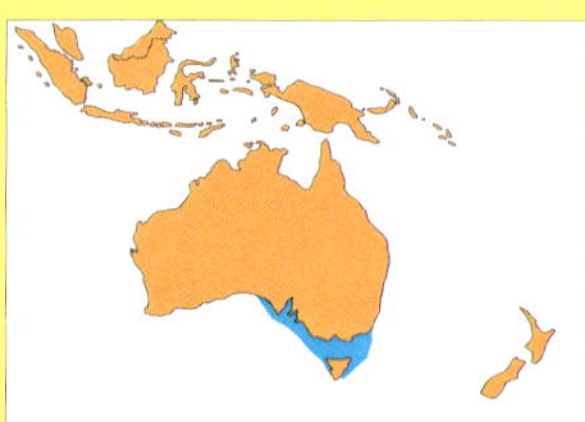

Common Sydney Octopus *Octopus tetricus*

This octopus reaches large sizes with an armspan of more than 2 metres. It is often seen by divers in the mouths of lairs, showing a white eye iris and orange arms. It feeds on shellfish, which it drills and poisons, then pulls the shells open and eats the paralysed contents. This species primarily emerges at night to hunt crustaceans and gather shellfish. It also preys on its own species. A closely-related species or subspecies occurs at similar latitudes in Western Australia.

The female lays tiny eggs (1–2 mm long) in strings on the roof of rock crevices or burrows, often blocking the entrance to the lair by holding rocks on its suckers to form a rock wall. The eggs hatch into small free-swimming young which are carried in the plankton.

Above: Lairs of *Octopus tetricus* can often be recognised by the bivalve shells scattered around their entrance. *Below:* It is easily identified in the field by its white eye. (Photos: M Norman)

IDENTIFICATION Typically grey to mottled brown with orange arm faces. Eye usually white. Skin consists of many small pavement-like patches and large papillae which can be raised over body to produce spiked appearance.

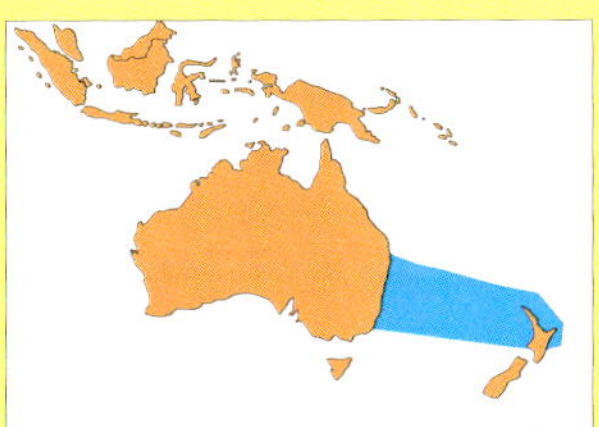

Club Pygmy Octopus *Octopus warringa*

Above: The male *Octopus warringa* has a modified club-like tip on third right arm. (Photo: D Paul) *Below:* A female guarding her eggs in kelp holdfast. (Photo: R Kuiter)

This small cryptic octopus reaches an armspan of 15 cm. It lives on rocky reefs and rubble areas and has been collected from intertidal pools to more than 100 metres deep. Virtually nothing is known of its biology and behaviour. It gets its name from the club-like tip on the modified sexual arm of the male. Females lay their eggs in the safety of holes deep in rocks and kelp holdfasts. The tiny eggs hatch into simple planktonic young which disperse in ocean currents, as far as New Zealand.

IDENTIFICATION Resting colour of mottled brown. Alarm pattern of black ring-like patterns on upper surfaces of mantle. Can also match purple or pink coralline algae. Male with swollen club tip on end of third right arm.

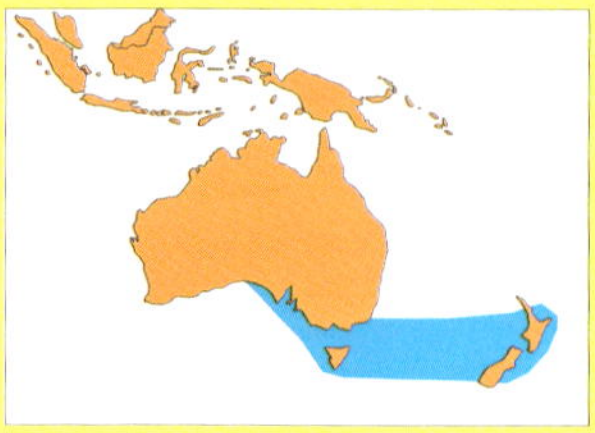

Eye-cross Octopus *Octopus* sp.

This octopus reaches an armspan of about 40 cm. It occurs on sand and rubble substrates to at least 100 metres. This species excavates a lair in rubble, blocking the entrance with loose pieces of rock or coral when at home. It only emerges at dusk or dawn to feed, collecting relatively passive prey such as shellfish (including nerites and mussels). The octopus drills the shells to poison and relax the occupants, so that it can extract and eat them. Established lairs are recognised by the large numbers of small shells around a blocked hole.

This octopus produces numerous small eggs and the tiny poorly developed young are likely to enter the plankton on hatching.

Above: The Eye-cross Octopus has a netmesh pattern formed by the dark grooves between regular patches. *Below:* Its common name comes from the four white bars around its eye. (Photos: M Norman)

IDENTIFICATION Common name derived from cross of four white bars around eye. Regular raised polygonal patches in skin separated by distinct darkened grooves. Produces very long sperm packages (spermatophores), about twice its mantle length.

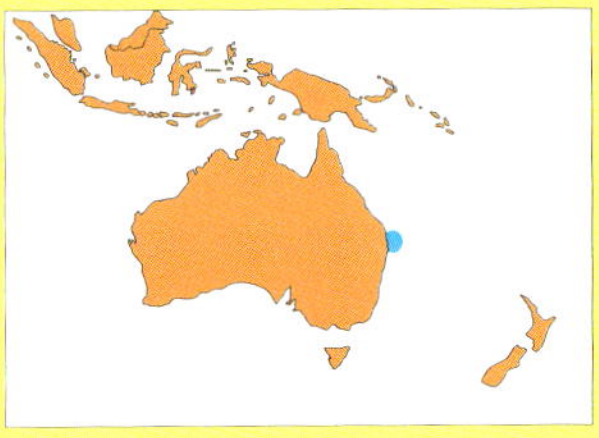

Day Drop-arm Octopus *Octopus* sp.

Above: The Day Drop-arm Octopus in alarm display with flared webs and dark circular markings. *Below:* The camouflaged colour pattern of the same individual. (Photos: A Semini)

This long-armed octopus has an armspan of about 20 cm. It occurs only on the intertidal coral reefs of a single island group on the southern Great Barrier Reef. It emerges during the day to forage for crabs while the tide is out, safe from the larger predatory fish which move over the reef during high tides. This octopus is an expert at camouflage and is extremely hard to find. If disturbed, it has two defense strategies. The first is to flare the webs between the arms and darken the colour pattern to look as large as possible (above). If the octopus is still attacked, it can drop one or more arms as wriggling decoys. Meanwhile it sneaks away and re-grows the lost arm(s).

This species mates from a distance, the male sending a long arm across rockpools to the female and shunting sperm packets along a gutter in this arm into the female's oviducts. Little else is known of the reproductive biology of this species.

IDENTIFICATION Mottled cream colour pattern, enlarged suckers on arms 2 and 3 of male, arms regenerating from base at same level (between 5th and 8th sucker). There are several related undescribed species elsewhere in northern Australia and Indonesia.

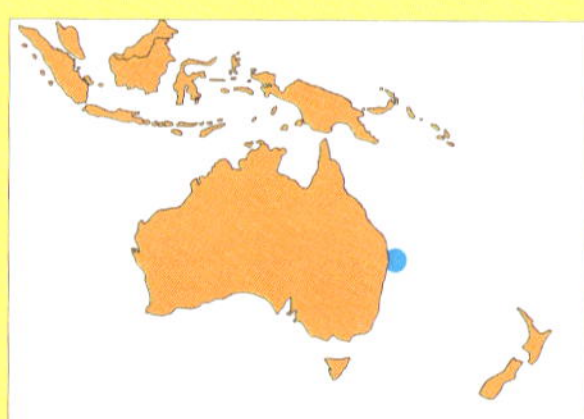

Mimic Octopus (undesribed species)

IDENTIFICATION May be related to Wunderpus, sharing long arms, banded webs and similar behaviours. More variable in colour pattern and has irregular white stripes along body. Eyes less stalked than in Wunderpus.

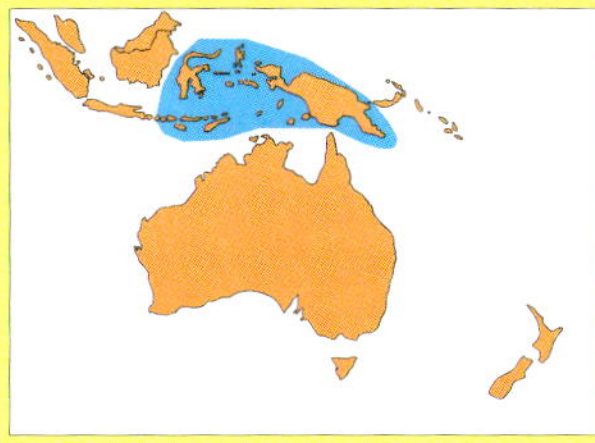

This species has very long narrow arms and can reach armspans of about 40 cm. It is found on sandy substrates in the shallow tropical waters of Indonesia where it often resides in the burrows of other animals. It is an expert at mimicking other animals, possibly as a way of deceiving potential predators or luring hapless prey. It has been observed to mimic flounders, anemones, sea snakes, mantis shrimps, lionfish and even a stingray. Not only does it look like the mimicked animal, it even behaves like them! The flounder mimicry includes rippling over the seafloor like the swimming fish.

The Mimic Octopus has been observed to feed on crabs. Little else is known of its biology with nothing known of its reproductive biology.

The Mimic Octopus can mimic a stingray complete with tail (*above right*) or a moving sea snake (*below*). (Photos: R Steene)

Wunderpus (undescribed species)

Above: Wunderpus's dramatic colour pattern may be impersonating poisonous animals such as the lion fish, or may advertise its own poisonous nature. *Below:* No other octopus has such stalked eyes as the Wunderpus. (Photos: A Kerstitch)

This spectacular long-armed octopus reaches an armspan of about 40 cm. It is known from shallow waters in the western central Pacific region from Vanuatu to Indonesia and the Philippines. It has mainly been found foraging over soft substrates. Individuals have been observed foraging in broad daylight, apparently undisturbed by passing fish. While foraging, the colour pattern is supplemented by a rapidly moving pulse of black up and down the eye stalks. There is still virtually nothing known of the biology of this undescribed species. Only a few animals have been observed in the wild.

IDENTIFICATION Superficially similar to Mimic Octopus but is easily distinguished by fixed pattern of distinct oval to elongate white patches on upper surface of mantle, against an even background of brown to red. Elongate stalked eyes form a Y-shape.

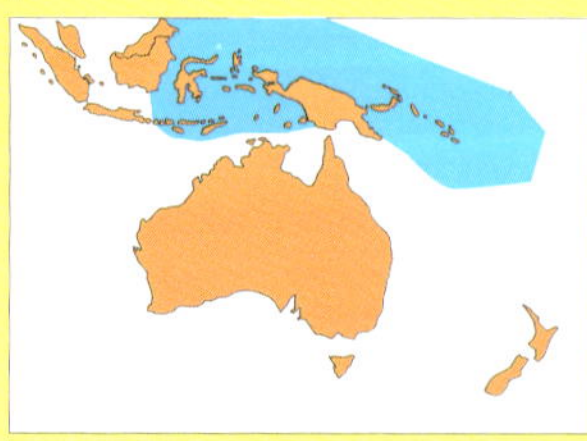

Argonauts or Paper Nautiluses (Family Argonautidae)

A juvenile female Southern Argonaut, *Argonauta nodosa*, about 5 cm across, has already laid down a shell to protect herself. The bumps on the shell are not so prominent in juveniles.

The shell of an adult *Argonauta nodosa*. Swarms of females occasionally wash ashore in their thousands in southern Australia. The species name comes from the bumps or nodules along the ridges on the side of the shell.

The shell of the Greater Argonaut, *Argonauta argo*, is about 30 cm across. This large argonaut lives in tropical and warm temperate waters around the world. The white shell is the finest in this group of octopuses, being narrow with low smooth ridges. The inner leading edges of the shell aperture are thickened and pigmentation is restricted to the first third of the crest. Little is known of the biology and life-cycle of this open-ocean octopus. Shells have washed ashore as far south as eastern Victoria.

The shell of the Lesser Argonaut, *Argonauta hians*, is approximately 10 cm across. This Argonaut is the smallest in the group. It occurs in tropical and warm temperate waters world-wide. The shell is cream to brown in colour, with raised ridges and widely-spaced bumps along the crest. As with *Argonauta argo*, nothing is known of its biology and life-cycle.

(Photos: D Paul)

Southern Argonaut *Argonauta nodosa*

Above: A swimming female *Argonauta nodosa* with the webs of its front arms spread over its shell. (Photo: R Kuiter) *Below:* A tiny male with modified arm curled within large pouch at the bottom. (Photo: B Reid)

Unlike the true chambered nautilus, the argonaut or 'paper nautilus' is a type of octopus. It spends its life in the surface waters of open oceans feeding on fish and shrimp. The female produces a paper-thin shell as a brood chamber for her eggs. She grows to about 30 cm in mantle length and lives her whole life within the shell, gripping it with her suckers. As she grows, she lays down her shell using the broad webs extending off her upper pair of arms. These webs are spread over the outside of the shell as she swims and feeds, using her funnel for jet propulsion.

The male argonaut is tiny (<3 cm) and has no shell but has a large modified arm in which he stores sperm. During mating, this arm breaks off in the female and the male dies. The sperm is then used to fertilise the eggs.

IDENTIFICATION *Argonauta nodosa* **has regular bumps (or nodes) within ridges across shell. Other species have smooth ridges across shell. Modified third arm of males in argonauts is on left hand side.**

Football Octopus *Ocythoe tuberculata*

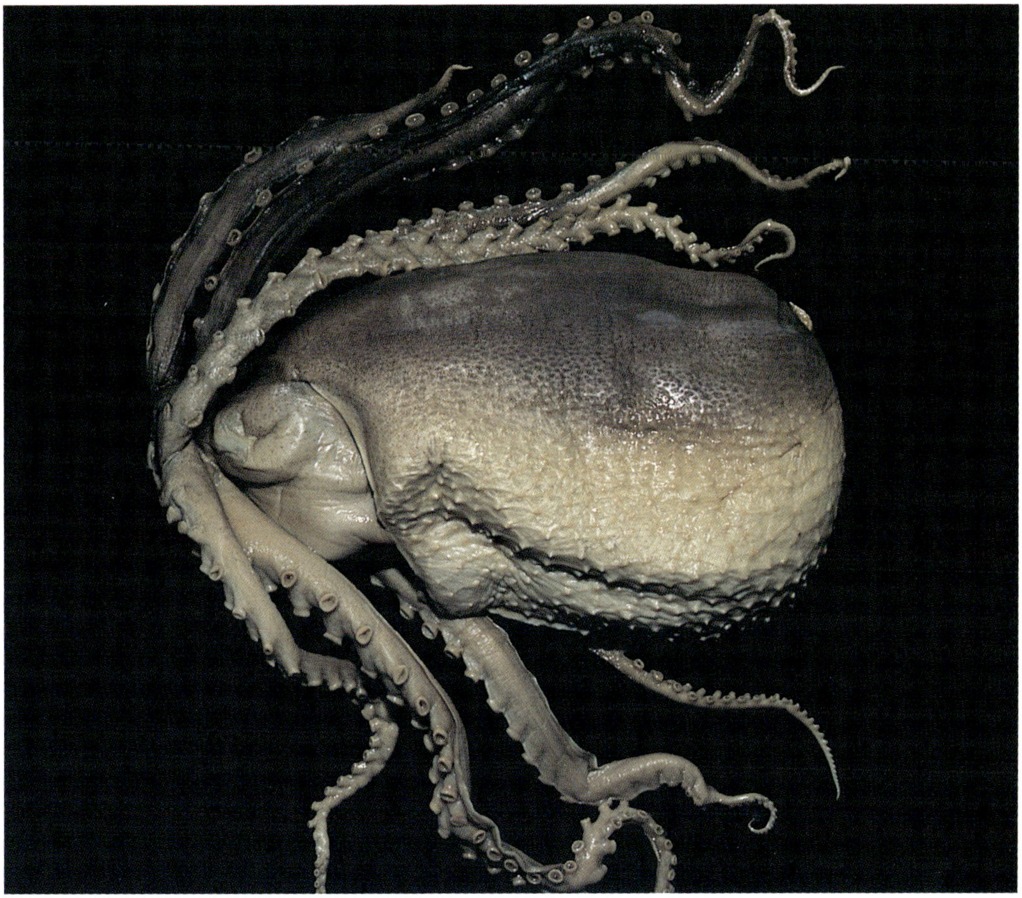

The female *Ocythoe tuberculata* has a mesh of ridges and tubercles on the underside of her mantle. (Photo: M Norman)

The females of this robust species have a body which reaches about 30 cm long. The underside of the mantle is sculptured with a mesh of ridges and tubercles. She is also the only cephalopod to possess a gas-filled swim bladder. The miniature male is only about 5 cm long and lacks the sculpture of the female. This species lives in open ocean from surface waters down to about 200 metres. Little is known of its behaviour. Males have been observed travelling inside hollow jellyfish (salps). When approached by divers, these males have jetted away from their refuge, later returning. The remains of this octopus have been found in the stomachs of many open ocean predators such as lancetfishes, tuna and dolphins.

The miniature male stores sperm in a specially modified third right arm which is detached and passed to the female during mating. The female broods her eggs within her long oviducts, thus making her the only live-bearing cephalopod in the world.

IDENTIFICATION
Females have netmesh sculpture underneath body. Males with modified third right arm which develops in pouch. Sperm is stored in base of this special arm.

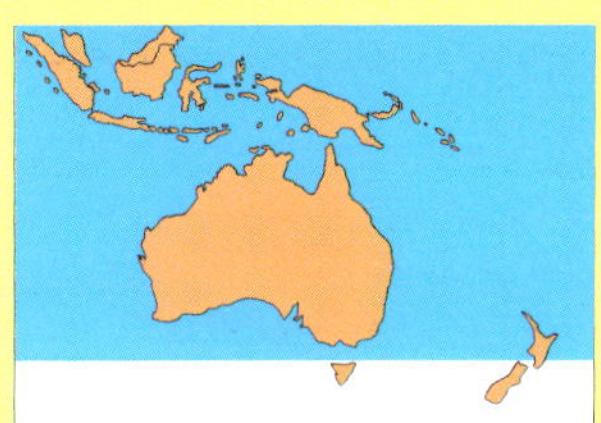

Blanket Octopus *Tremoctopus violaceus*

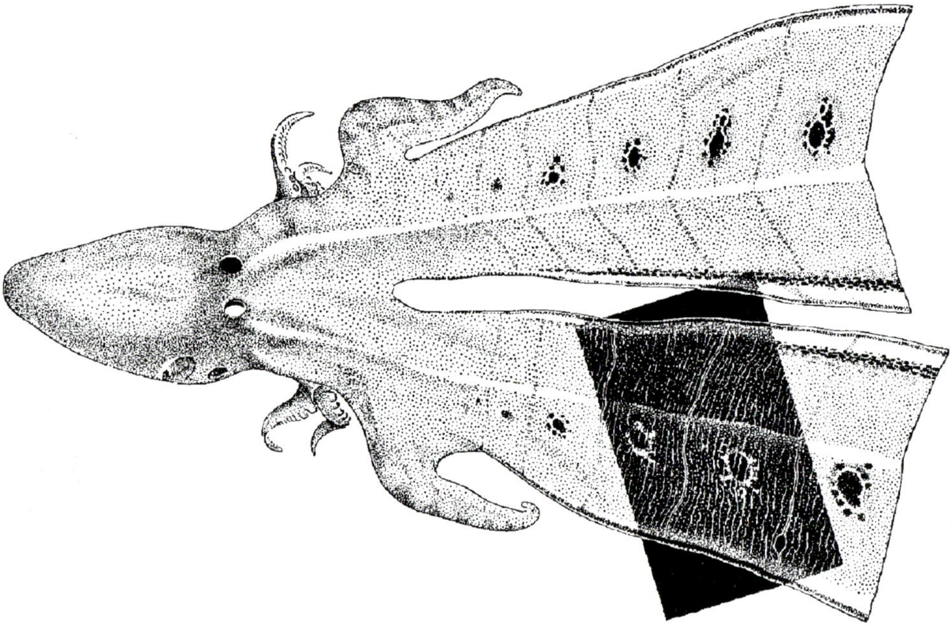

Above: Female *Tremoctopus violaceus* with long veil-like webs off upper arms. (Illustration after Portmann) *Below:* Head of female with water pores in dark upper surface. Eye is visible facing down within cream coloured layer. (Photo: M Norman)

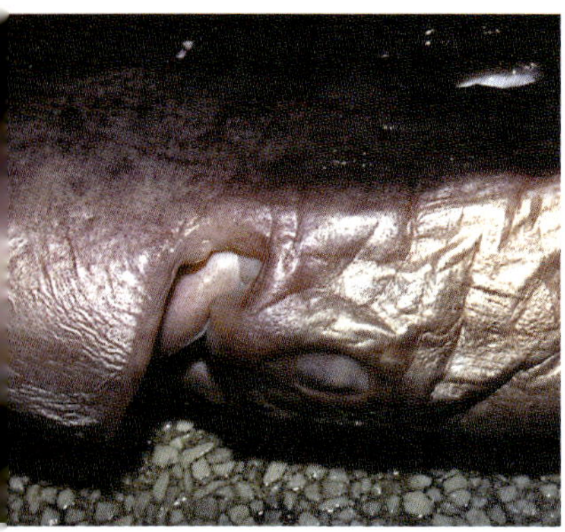

The female Blanket Octopus has long webs and reaches a total length of about 2 metres. The long thin webs trail behind the animal when jetting. By contrast, the miniature male has no webs and is only about 5–10 cm long. This unique animal lives in the open oceans of the world and little is known of its behaviour. The transparent webs of the front arms are segmented and can be shed like sheets of paper. The live animal has iridescent purple and gold spots within each web segment of unknown function.

The small male of this species has a specially modified third right arm in which sperm is stored. During mating the arm severs and crawls into the mantle of the female. The male then dies. Females carry the eggs until hatching, attached in strings to a small secreted rod. Juveniles have been found to arm themselves with the stinging tentacles of jellyfishes.

IDENTIFICATION Females are dark purple above and cream pink below, recognised by long thin webs. Males have large modified third right arm which develops in pouch. If opened, this arm has bulb-like sperm reservoir on tip.

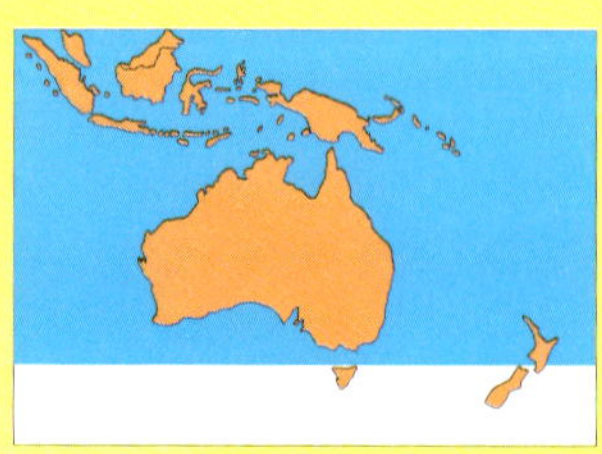

Cuttlebones – an identification guide

Many species of cuttlefish live in offshore waters at depths in excess of 100 m. The bones of dead animals are often found washed up on beaches. The biology and behaviour of most of these species remains largely unknown.

The bones on the following pages include both shallow and deep-water species and are likely to be found by beachcombers. For each species included, the dorsal, or upper, view of the cuttlebone is on the left and the ventral, or lower, view of the cuttlebone is on the right. In each case the approximate maximum length of the cuttlebone is given, followed by the distribution of the species. Cuttlebones may float on currents for considerable distances, so may be found outside the true range of a species.

Powdered cuttlebones have been used in the past as polishing powder. The powder has been added to toothpaste and used as an antacid or as an absorbent. Now the bones are used primarily as a calcium supplement for cage-birds, and carved to form jewellery moulds.

The cuttlebone of a cuttlefish lies within its mantle. The drawings below of the cuttlebone of *Sepiella inermis* show the things to look for when trying to identify a species.

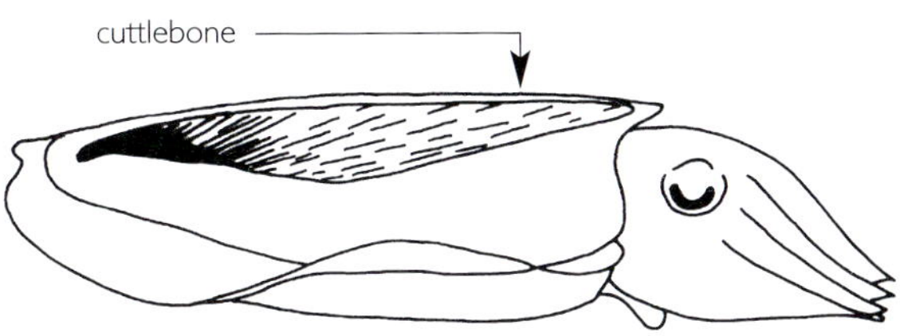

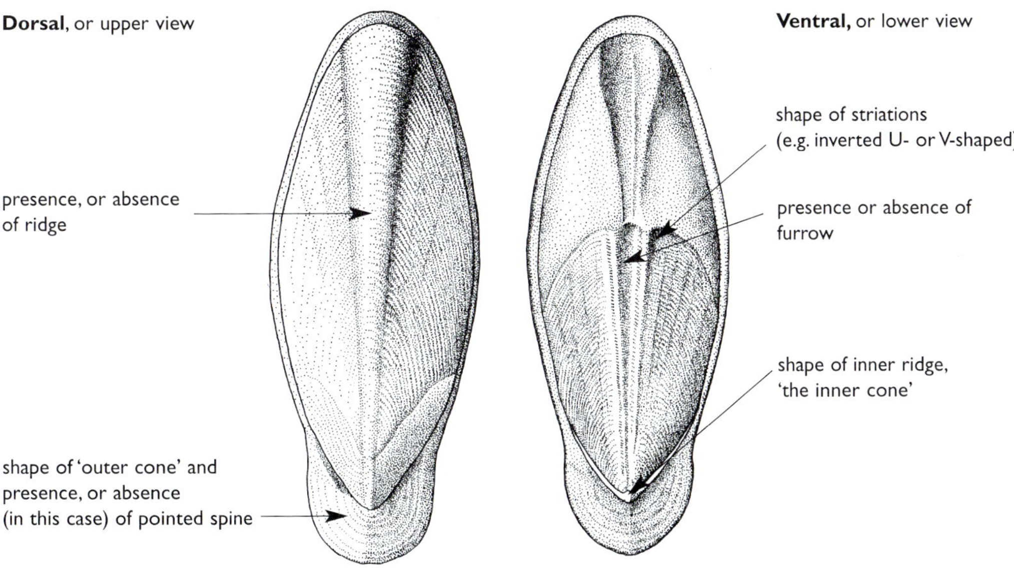

Sepia apama
50 cm. Southern Australia.

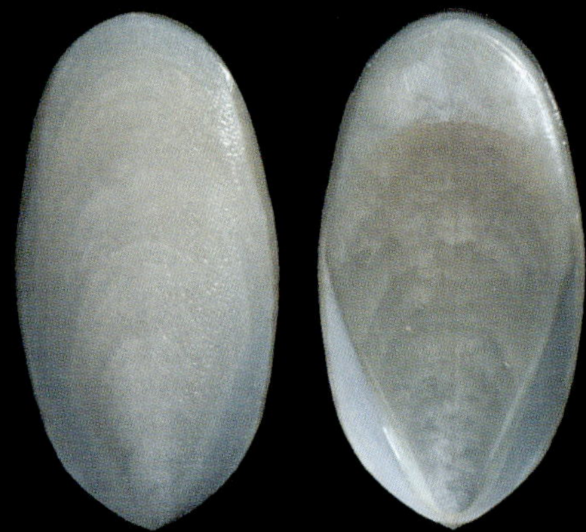

Sepia bandensis
5 cm. Philippines, Irian Jaya and the Marshall Islands.

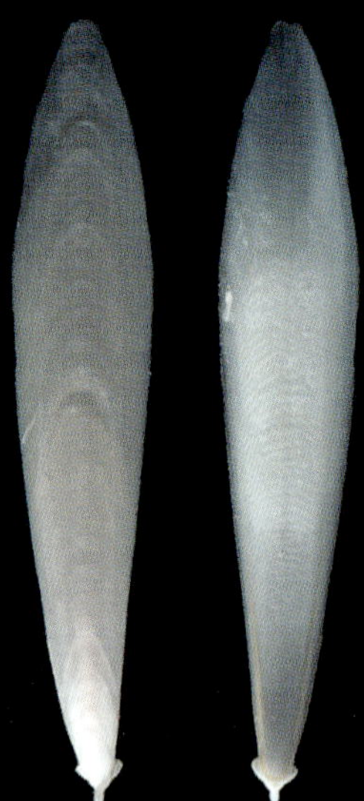

Sepia braggi
8 cm. Southern Australia, from southern New South Wales to Western Australia.

Sepia brevimana
10 cm. Widespread through south-east Asia, including the Java, Sulu and Celebes Seas.

Sepia chirotrema
8 cm. Southern Australia from Investigator Strait to Dirk Hartog Island, Western Australia.

Sepia cottoni
5 cm. Western Australia, from northwest of Broome to Rottnest Island.

Sepia cultrata
10 cm. Southern Australia.

Sepia elliptica
17 cm. Northern Australia and New Guinea, possibly as far north as the Philippines.

Sepia hedleyi
12 cm. Southern Australia.

Sepia irvingi
10 cm. Western Australia from the North West Shelf to Cockburn Sound.

Sepia koilados
6.8 cm. Western Australia, North West Shelf.

Sepia latimanus
50 cm. Northern Australia and widespread in the tropical Indo-West Pacific.

Sepia limata
4.2 cm. Eastern Australia, from southern Queensland to New South Wales.

Sepia mestus
14 cm. Eastern Australia and possibly elsewhere in the Indo West Pacific.

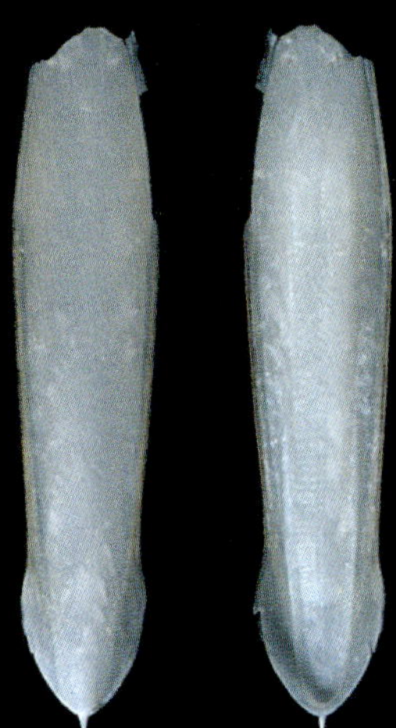

Sepia mira
4.4 cm. Eastern Australia, New South Wales from the Clarence River to off Wooli.

Sepia novaehollandiae
17 cm. Southern Australia from Shell Harbour, New South Wales to the North West Shelf, Western Australia.

Sepia opipara
15 cm. Northern Australia.

Sepia papuensis
11 cm. Widespread from the Philippines through Indonesia to northern Australia.

Sepia pharaonis
42 cm. Widespread through the Indo-West Pacific, to northern Australia.

Sepia plana
15 cm. Australia, North West Shelf.

Sepia plangon
14 cm. Eastern Australia.

Sepia rhoda
6 cm. Northern Australia from the Arafura Sea, Northern Territory to the North West Shelf, Western Australia.

Sepia rozella
14 cm. Eastern Australia.

Sepia smithi
17 cm. Timor, Arafura and Coral Seas to northern Australia.

Sepia senta
8.3 cm. Australia, North West Shelf.

Sepia sulcata
9.6 cm. Indonesia, Arafura Sea to Australia, North West Shelf.

Sepia whitleyana
17 cm. Eastern Australia from the Gulf of Carpentaria to New South Wales.

Metasepia pfefferi
8 cm. Irian Jaya and Papua New Guinea to northern Australia.

Sepiella mangkangunga
5.9 cm. Australia, Northern Territory.

Sepiella weberi
6.9 cm. Indonesia to northwestern Australia.

Note: Pictures of *Sepia bidhaia*, *Sepia recurvirostra*, *Sepia rex*, *Sepia sulcata* and *Sepia vercoi*, species that also occur in the region, are not included. See page 26 for a drawing of the *Sepia lycidas* cuttlebone.

References

ABRS (1998). *Mollusca: The Southern Synthesis, Part A.* (CSIRO Publishing: Melbourne.)

Carpenter, K.E., and Niem, V.H. (Eds) (1998). *FAO Species Identification Guide for Fishery Purposes. The living marine resources of the Western Central Pacific. Vol. 2. Cephalopods, crustaceans, holothurians and sharks.* (FAO: Rome.)

Clarke, M.R. and Trueman, E.R. (Eds) (1998). *The Mollusca Volume 12. Paleontology and Neontology of Cephalopods.* (Academic Press: London.)

Ellis, R. (1998). *The Search for the Giant Squid.* (The Lyons Press: New York.)

Hanlon, R.T. and Messenger, J.B. (1996). *Cephalopod Behaviour.* (Cambridge University Press: Cambridge.)

Lu, C.C. and Dunning, M.C. (1998). Class Cephalopoda. *In Mollusca: The Southern Synthesis, Part A.* (Eds P.L. Beesley, G.J.B. Ross and A.Wells.) pp. 451–563. (CSIRO Publishing: Melbourne.)

Nesis, K.N. (1987). *Cephalopods of the World.* (T.F.H. Publications: Neptune City, New Jersey.)

Norman, M.D. (1997). Australian octopuses. *Nature Australia*: 25(11), pp. 40–49.

Okutani, T. 1995. *Cuttlefish and Squids of the World in Colour.* (Okumura Printing Co.: Tokyo.)

Glossary

benthic living near the bottom of the sea
bivalve animal with a shell in two pairs hinged together (e.g. mussels and clams)
bycatch animals caught in addition to the target species
calcareous chalky, calcified by the inclusion of calcium salts (calcium carbonate)
chitin a horny, fingernail-like substance that forms the sucker rings, hooks and beaks of cephalopods
chromatophores muscular, pigment-filled sacs under the skin which are opened and closed in response to nerve action.
cirri long thin sensory papillae present between the suckers in deepsea finned octopuses and vampire squid
club tip of the tentacle, a pad with suckers
cuttlebone calcareous supporting plate in the upper side of the mantle of cuttlefish
fins a pair of muscular flaps on the sides of the mantle; used for locomotion, steering and stabilisation
funnel tube on the lower side through which water is expelled from the mantle cavity
gastropod a class of molluscs which includes snails, slugs and sea-hares, having a distinct head, with eyes and tentacles and often a single shell
gills paired structures, each made up of many leaf-like parts through which gas exchange takes place
gladius feather or rod-shaped supporting structure in squid and sepiolids (except cuttlefish)
haemocyanin copper-based blood pigment used to transport oxygen in the bloodstream — also responsible for the blue colour of the blood
hectocotylus the part, or parts, of the male arm(s) of cephalopods modified for transferring sperm to females
ink sac structure that stores the ink
iridocytes cells that reflect light to produce silvering and iridescence
keel a longitudinal ridge of skin along each side of the body found in some sand-dwelling octopuses
lens part of the eye which functions to produce an image on the retina (the light sensitive layer of the eye)
leucophores branched cells in the skin that scatter light of all wavelengths
light organ a structure that produces bioluminescence — may be present in the eyes, embedded in the mantle muscle, head, arms, tentacles, or on the viscera
mantle the spherical or tubular body of a cephalopod, which contains all the body organs
medusoid swimming when octopuses swim by expanding their arms and webs and closing them together like the pulsing bell of a jellyfish
ocellus dark false-eye spots found in 'ocellate' octopuses, one spot below each eye between the bases of the second and third arms
papillae small elongate digits of skin
photophores organs that produce bioluminescent light to eliminate the silhouette of the cephalopod against daylight coming down from the surface
posterior situated behind or the hinder part of the body
radula the toothed tongue of cephalopods
rhynchoteuthion the juvenile stage of certain squids where the tentacles are fused to form an elongated tube
spermatophores packets of sperm
spine sharp, pointed extension on the posterior end of the cuttlebone
suckers muscular suction-cups on the arms, tentacles, and sometimes around the mouth
tentacles elongate, stalked appendages used to capture the prey
tetrodotoxin a strong nerve toxin produced by bacteria in the salivary glands of blue-ringed octopuses
umbilicus the innermost coils of a nautilus shell, often looking like a belly button
web a membranous sheet that extends between the arms of many octopuses

Index